In this tremendously thoughtful and very engaging book, Patrycja Sasnal offers a brilliant study of two of the most important political theorists of the twentieth century, Fanon and Arendt, and then interprets violence in the contemporary Arab world using their tools of analysis. Her study is original, insightful, and challenging.

—*Susannah Heschel, Eli Black Professor of Jewish Studies, Dartmouth College*

Arendt, Fanon and Political Violence in Islam

This book looks at contemporary political violence, in the form of jihadism, through the lens of a philosophical polemic between Hannah Arendt and Frantz Fanon: intellectual representatives of the global north and global south.

It explores the relationship of Arendt's thought, mostly as expressed in *On Violence* (1969), to Fanon's *The Wretched of the Earth* (1961) and the transposition of that relationship to the contemporary phenomenon of violent Islamic extremism. The book reveals a greater commonality between Fanon and Arendt as well as the universal function of jihadism that satisfies the conditions for political violence, as categorized by Fanon in the global south and Arendt in the global north. Read in tandem, Arendt and Fanon help uncover the fundamental problems of our European, American, Middle Eastern and African political systems as well as north-south relations. By studying political theory, the book finds global political commonalities in a postcolonial reality.

Written in an accessible style, this book will be of great interest to undergraduates and graduates in philosophy, political sciences and international relations (IR), sociology and Middle Eastern studies as well as scholars and professionals interested in radicalization; violent extremism; and the foreign policies of European, Middle Eastern and African countries.

Patrycja Sasnal is a political scientist, philosopher and Arabist specializing in IR in the Middle East, with a focus on radicalization, political violence, postcolonial theory and migration. She is currently the head of the Middle East and Africa programme at the Polish Institute of International Affairs.

Routledge Research on Decoloniality and New Postcolonialisms

Series Editor: Mark Jackson, Senior Lecturer in Postcolonial Geographies, School of Geographical Sciences, University of Bristol, UK

This series provides a forum for innovative, critical research into the changing contexts, emerging potentials, and contemporary challenges ongoing within postcolonial studies. Postcolonial studies across the social sciences and humanities are in a period of transition and innovation. From environmental and ecological politics, to the development of new theoretical and methodological frameworks in posthumanisms, ontology, and relational ethics, to decolonizing efforts against expanding imperialisms, enclosures, and global violences against people and place, postcolonial studies are never more relevant and, at the same time, challenged. This series draws into focus emerging transdisciplinary conversations that engage key debates about how new postcolonial landscapes and new empirical and conceptual terrains are changing the legacies, scope, and responsibilities of decolonising critique.

Coloniality, Ontology, and the Question of the Posthuman
Edited by Mark Jackson

Unsettling Eurocentrism in the Westernized University
Edited by Julie Cupples and Ramón Grosfoguel

History, Imperialism, Critique
New Essays in World Literature
Edited by Asher Ghaffar

Arendt, Fanon and Political Violence in Islam
Patrycja Sasnal

For more information about this series, please visit: https://www.routledge.com/Routledge-Research-on-Decoloniality-and-New-Postcolonialisms/book-series/RRNP

Arendt, Fanon and Political Violence in Islam

Patrycja Sasnal

Routledge
Taylor & Francis Group
LONDON AND NEW YORK

First published 2020 by Routledge

2 Park Square, Milton Park, Abingdon, Oxon, OX14 4RN
605 Third Avenue, New York, NY 10017

Routledge is an imprint of the Taylor & Francis Group, an informa business

First issued in paperback 2020

British Library Cataloguing-in-Publication Data
A catalogue record for this book is available from the British Library

Library of Congress Cataloging-in-Publication Data
A catalog record has been requested for this book

ISBN: 978-0-367-25959-4 (hbk)
ISBN: 978-0-367-78790-5 (pbk)

Typeset in Times New Roman
by codeMantra

To Marek

Contents

Illustrations

Figure

Tables

Acknowledgements

I am indebted to Magda Środa from the University of Warsaw, who introduced me to Hannah Arendt and made me properly structure the topic of what was going to be a thesis in philosophy. Magda and Agnieszka Nogal were the first reviewers of the very first draft of this book. If it wasn't for their welcome reception, most probably I would not have worked on this text more.

I cannot thank enough the reviewers commissioned by Faye Leerink at Taylor & Francis and Faye herself. They helped improve the text significantly, although – of course – I am solely to blame for its shortcomings.

My husband, Marek, encouraged me to work on this book daily by asking the same, usually annoying, question whenever he saw me at the desk: "Are you working on the book?" I want to thank him for that dearly.

Abbreviations

BPF	Between Past and Future
BSWM	Black Skin White Masks
EiJ	Eichmann in Jerusalem
HC	Human Condition
OoT	Origins of Totalitarianism
OR	On Revolution
OV	On Violence
WotE	Wretched of the Earth
WR	We Refugees

Introduction

Violence and Europe-Middle East coupling 1960s–2019

This study looks at contemporary political violence in the form of jihadism through the lens of a philosophical polemic between Hannah Arendt and Frantz Fanon: intellectual representatives of the global north and global south. It does not aspire to analyse political violence in Islam comprehensively, neither in doctrine nor praxis. Its interest is (1) the relation of Hannah Arendt's thought mostly as expressed in *On Violence* (OV, 1969) to Frantz Fanon's *The Wretched of the Earth* (WotE, 1961) and (2) the transposition of that relation to the phenomenon of violent Islamic extremism. Throughout the study a greater commonality between Fanon and Arendt is revealed than in existing literature, as well as the universal function of jihadism that satisfies the conditions for political violence as categorised by Fanon in the global south and Arendt in the global north.

Fanon's views on violence and Arendt's polemics with them are as fresh and relevant a reading today as they were half a century ago. It is not a proper philosophical discussion – Arendt's OV, in which she lays out her views polemicising with Fanon's WotE, was published eight years after the latter's publication and Fanon's death. Yet their two works draw a single line of thought. It starts with Fanon's supposition that through violence the colonised will liberate himself and create new humanity for the world. It continues in Arendt's developing this idea into a comprehensive lecture on the relation of violence to power and, more generally, on the state of politics. Violence at campuses in the US and student revolts in Europe inspired Arendt to address the topic. Earlier, colonial violence and counterviolence in Algeria made Fanon write WotE. Today's terrorist violence invites a rereading of the debate between Arendt and Fanon – a reading that could culminate in thoughts on the political systems in the north and south.

Contrary to the times of Fanon and Arendt when violence was prevalent, since 1989 the Western world has nominally enjoyed more than two decades of "peace." Formally, indeed, the cold war ended and the "third wave" of democratisation[1] swept Central Europe and parts of Asia. The dominant narrative had it that we lived in times of unprecedented peace, end of history, etc. Several of the Western world projects validated such opinions: the European Union as an unparalleled political and ideological enterprise brought and secured European peace. The cherry on the cake were findings of Steven Pinker, a Harvard professor in psychology, who claimed that as human beings we are getting more benign and so there is less violence.[2] Even though he can be faulted with i.a. ignoring less ostentatious forms of violence, the praise he has won for his book outweighed the criticism.[3]

But for many non-Europeans, many Arabs, the year 1989 was no different from any other year and certainly not in that it reduced violence. In 1990 Saddam Husain annexed Kuwait, which triggered the second Gulf War with the United States invading Iraq, a country that had barely ended a bloody war with Iran (1980–88). Among the few Middle Eastern rulers who opposed American intervention then was Yasser Arafat, the Palestinian leader whose compatriots had been uprising against Israel in the first Intifada since 1987. In 1991 after the first democratic elections in Algeria, which poised the Islamists to win majority of seats in the parliament and form the government, the army seized power and a civil war began that took the lives of between 100,000 and 200,000 people, known since as *la décenie noire*. In that decade the US and UK invaded Iraq in 1998 in the operation Desert Fox and Israel raided Hezbollah in Lebanon in 1993 and 1996. Despite this violence the 1990s were to be remembered as the time when Israelis and Arabs began negotiating agreements – though that memory stuck only in the Western mind.

The Middle East experienced violence – inter and intrastate violence – and so it remembered violence, not peace. In 2001 the US invaded Afghanistan, where the perpetrators of the terrorist attack on the World Trade Center in New York were hiding. The war continues until today. It has claimed more than 100,000 victims.[4] In 2003, under the pretext of Iraq possessing weapons of mass destruction and giving Al-Qaeda safe haven the US invaded Iraq again. The intervention has cost the lives of 270,000 people officially, and more than 650,000 unofficially.[5] Alongside foreign interventions foreign-inspired and proxy wars Middle Eastern governments – though none of these institutions deserves that name – have jailed, tortured or otherwise oppressed millions of people.

Most recently the war in Syria turned out to cause the greatest humanitarian crisis of the 21st century – it took the lives of half a million people, predominantly at the hands of the brutal government in Damascus, and displaced, internally and externally, more than

10 million Syrians. The region is home to most refugees in the world. The numbers, even though they cannot be beaten by statistics from any other region, still say nothing about the scale of human suffering. Only when the UN declared the war in Syria the greatest humanitarian crisis the war in Yemen flared up, as a result of which 16 million people wake up hungry every day in 2019.[6] Each person killed, disabled, suffering from hunger or forced out of their homes, is a story of grief for at least a few, although usually many more of its family members, given the size of the families. The grief and frustration radiate throughout the region – a region that has become the bleeding wound of humanity.

The Middle East is plagued by violence, the terror kind of violence that destroys power and citizenship in Arendtian terms, and also one that is instigated from outside of the region, often by former colonies or by governments allied with Western powers. The post-cold war peace never reached the Middle East or, often, came at the expense of its people – the talk of a democratic wave or peaceful end of history are unknown concepts that smell of hypocrisy. The hypocrisy is an interesting phenomenon of, lexically, "claiming to have higher standards or more noble beliefs than is the case."[7] Etymologically, it is a way of playing on stage, behind a mask (gr. *hypokrisis*). What else if not a play is it on the part of Europe, who like Shakespeare's Claudius claims to have little to do with the violence outside of its territory, yet it never recognised having been its agent, "a net exporter of violence."[8] More so, the violence that bludgeoned more than 50 million Europeans between 1914 and 1945 "had been developed and practiced previously outside Europe."[9] Violence is a boomerang. If it ravages the Middle East, it will come home to roost.

* * *

In October 2017, Outoudert Abrous, the editor-in-chief of the leading Algerian daily "Liberté" was asked, on a trip to Europe and upon introduction, if he preferred to speak French or Arabic. His hosts could speak both. He replied that he could not speak Arabic, that he was Algerian and so he wanted to speak French. His response was not uncommon: in Algeria, Tunisia, Lebanon and Morocco many Arabs prefer speaking French, particularly with foreigners, as Arabic is considered inferior to European languages and, certainly, inferior to the language these Arabs learnt at schools: the French. Abrous must have exaggerated. Most probably he could speak Arabic but did not think it prestigious enough to let it be known. On a different occasion a well-known Moroccan was attending an International Relations conference in Ankara, Turkey. Having been served tea, he commented: "Unlike Arabs, Turks are civilized." The tea was without sugar – it is often served with sugar in the Arab world. During that afternoon the diplomat made several other comments that revealed his disdain for Islam and "uncivilized" Arabic traditionalism.

These encounters show how unresolved and undigested the relationship between the local and colonist's culture is in the Middle East. It epitomises the class-like societies in the Arab world, where some can only speak Arabic and can never travel abroad, while others do not want to speak any other language than the language of the former colonist, revel in its culture and abhor the one at home. It reveals how easily and quickly Europe has forgiven itself for the past that spasmodically lets itself known until today.

Colonialism must have a different pace from other political processes – like trees it grows slowly. It requires constructing lives in space and time – people erect houses, join them up in settlements, found schools, start working in district convenience stores, make friends with the people next door. Children are born and go to school, make friends again, a form of political life grows roots in a place where another form of political life had been sown (or – as we shall see – maybe none existed). It takes time for such a form of political life to grow. Once it's developed into a tree with strong rhizomes it does not suffice to cut it – even to the ground – the stems will still remain buried. With time they begin to bud again and grow. In the Middle East the residue of colonialism lingers on in stark inequalities, state and non-state violence and xenophobia. A descendant of the colonial bourgeoisie is there in the form of the Arab elite, as tired and inactive as ever. The remaining colonialism shows not only and not primarily in the global system under the mask of US foreign policy or globalisation and its institutions as Homi Bhabha pointed[10] in his introduction to the new translation of WotE. The inscrutable residue of colonialism remains in the rigid class feature of Arab societies, in the betrayal of the elite and the political class, perhaps in the absence of politics altogether (what most probably Hannah Arendt would point to). The presence of a colonist who is in fact often a local – a black skin into which the white mask has grown – is more oppressive because of its local, concealed and yet onerous nature. A 100-year-old Schindler elevator in a run-down hotel in *wust al-balad* – central Cairo – rouses nostalgy for a time truer, more straightforward, when differentiating the oppressor from the oppressed was easier, albeit more stereotypical.

The world of today resembles that of the 1960s, at least as read in the fear in Fanon's WotE and Arendt's OV. The feeling of injustice permeates the Middle East and, increasingly, the West, violence endures and reappears in Europe where it had seemed a thing of the past. The responsibility for injustice and violence is illusive and vague – it is being flipped between voluminous terms such as "Islam" or "the West" that abstractly identify the wrongdoer to provide a false sense of self-assurance. The debate between Arendt and Fanon helps uncover the fundamental problems of our, European and Middle Eastern, political systems and relations between them.

Notes on literature, purpose and structure

Arendt's polemic with Fanon is the subject of several scholarly works. An article by Elizabeth Frazer and Kimberly Hutchings from 2008 is the only one that deals with the polemic separately and par excellence.[11] A chapter in the 2014 Kathryn Gines's book *Hannah Arendt and the Negro Question* covers Arendt's criticism of Fanon.[12] Gines is also the author of a chapter in *Phenomenologies of Violence* (edited by Michael Staudigl and published by Brill in 2013) titled *Arendt's Violence/Power Distinction and Sartre's Violence/Counter-Violence Distinction: The Phenomenology of Violence in Colonial and Post-Colonial Context.*[13] She mentions the polemic in question, although she focuses on the differences between phenomenological takes on violence by Arendt and Sartre, without a direct reference to the topic of this book.

WotE inspired an ocean of literature, specifically since the 1980s and most recently there is a resurgence of volumes about Fanon himself.[14] Even though Arendt continues to stimulate an equally great number of minds, the literature specifically on OV is relatively rarer, since OR seems to be more popular a source of Arendt's thoughts about violence.[15] The one that stands out is Christopher J. Finlay's article "Hannah Arendt's Critique of Violence," in which Finlay reconstructs Arendt's take on violence based on OV and OR to show if and how it differs from George Sorel's and – to a lesser extent – Fanon's understandings of violence.[16] Another noteworthy position is Ned Curthoys's chapter analysing OV in the context of Algerian decolonisation.[17] A magnificent work about Arendt's thought and International Relations by Patricia Owens[18] covers Arendt's views on political violence in their essence but does not absorb Fanon's thinking enough – neither is it its ambition – to juice the essence of their commonality.

Their differences of opinion are noted in several other works that focus on either of the philosophers or on political violence. Mark Muhannad Ayyash's article on the paradox of political violence refers to Arendt's and Fanon's views on violence.[19] A Richard J. Bernstein book of 2013 on violence, published by Polity, separately looks at the thoughts on violence put forward by Carl Schmitt, Walter

Benjamin, Arendt, Fanon and Jan Assmann without, however, giving the Fanon-Arendt couple a greater attention.[20] Simon Swift's book "Hannah Arendt" has a section about Fanon, in which both philosophers' thoughts are discussed, although principally on identity.[21] Swift has also written, convincingly, about Arendt's engagement on violence and vitalism with Benjamin's and Bergson's writings.[22] Of importance in this study is Annabel Herzog's understanding of Arendt's position on violence, mostly because she persuasively highlights the connection between violence and the public realm,[23] as well as Herzog's elucidation on Arendt's take on responsibility.[24]

Apart from these works that looked, even if *en passant*, at the thoughts on violence by Arendt and Fanon, there have been several that discussed political violence in Islam in its generality, with or without a scholarly background.[25] Most of these works looked at violence as a specifically "southern" phenomenon, contrary to the perspective acquired in this book. The aforementioned works have all been published prior to the ISIS's takeover of parts of Syria and Iraq in 2014, and the terrorist attacks in Europe since 2014. After these events and with almost 40,000 foreign fighters in the ranks of jihadism in 2016 the understanding of the global power of this ideology had to change. One may argue that since a terrorist organisation was able to form a pseudo-state Talal Asad's sobering reminder from *On Suicide Bombing* has lost some of its power:

> I am simply impressed by the fact that modern states are able to destroy and disrupt life more easily and on a much grander scale than ever before and that terrorists cannot reach this capability.[26]

This book will not fill the void left by absence of work confronting Arendt with Fanon, however useful it would be, but it will, first, rethink their relation, and then transmute Fanon's and Arendt's arguments about violence into a contemporary debate about jihadi or Islamic terrorism in the 21st century, a debate that is also revealing about the state of politics in today's world. By doing this the study wants to achieve three goals:

1 **Take a novel approach to Hannah Arendt's polemic with Frantz Fanon about political violence and power.** In literature Arendt and Fanon are presented as being on opposite ends with their attitudes to political violence. The book argues that neither did Fanon advocate unconditional violence, nor was Arendt unconditionally against all forms of political violence. This study wants to deradicalise Fanon, while radicalising Arendt. Both are read

through the other's readings and both, I argue, can be interpreted as philosophers of north-south unity in their own right. Fanon, in particular, is often portrayed as the intellectual father of the anti-white and anti-Western/anti-Northern violence, including jihadist violence. The book attempts to disprove that notion.

2 **Classify reasons for the appeal of contemporary jihadist violence** (as the most common form of non-state political violence) **to show its global universality.** It lists factors that push to radicalisation, based on specialised literature, and then finds the same reasons in Arendt's and Fanon's writings. Their criticisms of the public spaces in the north and south share similarities and are as valid today as in the 1960s. Eventually, a conclusion can be drawn that Fanon is as much a philosopher of the north as Arendt is of the south.

3 **By reading political theory find and discuss global political commonalities in a postcolonial reality.** The analysis consists of a theoretical element (a polemic between emblematic philosophers of the north and south) and a practical element (political violence perpetrated by Muslims). Fanon and Arendt called for and analysed applications of political theory – this study has a similar goal: by finding similarities in theory (the theories of Fanon's and Arendt's) it wants to reveal them in practice (global jihadi violence).

"Political violence in Islam" is a broad category, encompassing both state and non-state violence, perpetrated in the Muslim world or by Muslims. In this book only violence in Islam in the form of jihadism – a non-state violence that cannot exist without state violence – will be singled out for two reasons. Firstly, unlike many other forms of contemporary violence, jihadism can be related to political conditions in the north and south and to philosophies of both Fanon and Arendt. The ideology is considered to culturally pertain to the Middle East, but it draws Middle Easterners and Westerners alike. Literature shows that Western converts to Islam are overrepresented in extremist groups.[27] Jihadism has turned out to be a convenient response to political problems both in the north and south, even though these problems are not identical. Certainly, other kinds of extremisms and radicalisation in general can result from those problems as well but no other contemporary violent ideology is as global and as popular as jihadism. Secondly, this kind of violence is of symbolic value in this study: in a single phenomenon of jihadism the thoughts of Fanon and Arendt find their validation. The causes for the appeal of jihadism are nothing else than the conditions for violence found in both philosopher's writings. In this way reality has connected Fanon and Arendt before scholarship could.

Although violence in Islam in its generality is not the primary topic of this book (its particular form serves as a case study here) it necessitates a clarification: there is as much political violence in Islam as there is in a human being.[28] Religious tones only give a radicalised individual a stage to what would catalyse otherwise or at a different time. The choice of the stage is not accidental: the revolutionary and utopian jihadism is better tailored to radicalisation in a globalised world than anarchism, communism or fascism. Radicalisation is a commonplace socio-political phenomenon that has been with humanity since time immemorial, regardless of culture, religion, language or race. More thorough engagement with the topic can be found in Chapter 4, while here it suffices to rest on the words of the renowned French Orientalist Maxime Rodinson from his *Europe and the Mystique of Islam* of 1988:

> A knowledge of Islam and the images of Islam, particularly in these times, could be an important key to the understanding of this world. There are many people who are now afraid of Islam. It is terribly true that many frightening acts are committed in the name of Islam, but these are no worse than those committed in the names of Christianity, Judaism, Freedom, and so on. Islamic peoples form a part of the world's underprivileged masses. They quite naturally long to improve their situation and will employ any means, right or wrong, to achieve that goal. This is a fundamental rule of all human nature.[29]

In the first chapter Fanon and Arendt are presented in tandem with an emphasis on the philosophies that influenced them in the way that later resounded in WotE and OV. In the chapter "Violence vs. Power" the thoughts on violence in WotE and OV are discussed in separation because indeed on the textual level and, partially, interpretative too they were not consonant. A short summary of the famed and notorious preface to WotE by Jean Paul Sartre precedes that part. The commonality between Fanon and Arendt comes to view in the chapter on the "New Humanism," which discusses their philosophy of responsibility and commitment, the political creation of man and their understanding of a nation. In the final part of the study jihadism – the most conspicuous manifestation of modern political violence – is analysed through an overview of material reasons for jihadist appeal.[30] The origins of modern violence, as exposed by jihadi terrorism and the "Islamisation of radicalism" (as opposed to radicalisation of Islam) become discernible through the prism of the Fanon-Arendt views.

Notes

1 Samuel P. Huntington, *The Third Wave: Democratization in the Late Twentieth Century* (Norman: University of Oklahoma Press, 1993).
2 Steven Pinker, *The Better Angels of Our Nature: Why Violence Has Declined* (New York: Viking, 2011).
3 For critical reception see: John Gray, "Steven Pinker Is Wrong about Violence and War," *The Guardian*, March 13, 2015, accessed March 28, 2019, www.theguardian.com/books/2015/mar/13/john-gray-steven-pinker-wrong-violence-war-declining.
4 "Afghan Civilians | Costs of War," Watson Institute of International and Public Affairs, Brown University, accessed December 1, 2017, http://watson.brown.edu/costsofwar/costs/human/civilians/afghan.
5 Tim Parsons and JH Bloomberg School of Public Health, "Updated Iraq Survey Affirms Earlier Mortality Estimates," Johns Hopkins Bloomberg School of Public Health, accessed December 1, 2017, www.jhsph.edu/news/news-releases/2006/burnham-iraq-2006.html.
6 "Yemen Emergency," *World Food Programme*, accessed March 21, 2019, www1.wfp.org/emergencies/yemen-emergency.
7 "Hypocrisy | Definition of Hypocrisy in English by Oxford Dictionaries," *Oxford Dictionaries | English*, https://en.oxforddictionaries.com/definition/hypocrisy.
8 Donald Bloxham and Robert Gerwarth, eds., *Political Violence in Twentieth-Century Europe* (Cambridge: Cambridge University Press, 2011), 6.
9 Bloxham and Gerwarth, 5–7. Arendt herself develops such assumption in OoT.
10 For interesting polemics with Bhabha see Nigel Gibson, "Is Fanon Relevant? Toward an Alternative Foreword to 'The Damned of the Earth,'" *Human Architecture: Journal of the Sociology of Self-Knowledge* 5, no. 3 (June 21, 2007).
11 Elizabeth Frazer and Kimberly Hutchings, "On Politics and Violence: Arendt Contra Fanon," *Contemporary Political Theory* 7, no. 1 (February 1, 2008): 90–108. The article develops a thesis that although in some respects Arendt and Fanon complement each other, Arendt's thoughts on violence are "less insightful" than Fanon's.
12 Kathryn T. Gines, *Hannah Arendt and the Negro Question* (Bloomington: Indiana University Press, 2014).
13 Kathryn Gines, "Arendt's Violence/Power Distinction and Sartre's Violence/Counter-Violence Distinction. The Phenomenology of Violence in Colonial and Post-Colonial Context." In *Phenomenologies of Violence*, ed. Michael Staudigl (Leiden: Brill, 2013), 123–44.
14 For examples of recent works see: Leo Zeilig, *Frantz Fanon: The Militant Philosopher of Third World Revolution* (London; New York: I.B. Tauris, 2016); Peter Hudis, *Frantz Fanon: Philosopher of the Barricades* (London: Pluto Press, 2015); Christopher J. Lee, *Frantz Fanon: Toward a Revolutionary Humanism* (Athens: Ohio University Press, 2015).
15 See: Patricia Owens, *Between War and Politics: International Relations and the Thought of Hannah Arendt* (Oxford: Oxford University Press, 2007); Bat-Ami Bar On, *The Subject of Violence: Arendtean Exercises in*

Understanding (Lanham: Rowman & Littlefield, 2002); Craig J. Calhoun and John McGowan, *Hannah Arendt and the Meaning of Politics* (Minneapolis: University of Minnesota Press, 1997).

16 Christopher J. Finlay, "Hannah Arendt's Critique of Violence," *Thesis Eleven* 97, no. 1 (May 2009): 26–45.

17 Ned Curthoys, "The Refractory Legacy of Algerian Decolonization: Revisiting Arendt on Violence," in *Hannah Arendt and the Uses of History. Imperialism, Nation, Race, and Genocide*, ed. Richard King and Dan Stone (New York, Oxford: Berghahn Books, 2007), 109–29.

18 Owens, *Between War and Politics*.

19 Mark Muhannad Ayyash, "The Paradox of Political Violence," *European Journal of Social Theory* 16, no. 3 (August 1, 2013): 342–56.

20 Richard J. Bernstein, *Violence: Thinking without Banisters* (Cambridge, UK: Polity, 2013).

21 Simon Swift, *Hannah Arendt* (London; New York: Routledge, 2009).

22 Simon Swift, "Hannah Arendt, Violence and Vitality," *European Journal of Social Theory* 16, no. 3 (August 2013): 357–76.

23 Annabel Herzog, "The Concept of Violence in the Work of Hannah Arendt," *Continental Philosophy Review* 50, no. 2 (2017): 165–79.

24 Annabel Herzog, "Hannah Arendt's Concept of Responsibility," *Studies in Social and Political Thought* 10 (August 2004): 39–56.

25 Jarret M. Brachman, *Global Jihadism: Theory and Practice* (London; New York: Routledge, 2008); Eyal Weizman, *The Least of All Possible Evils: Humanitarian Violence from Arendt to Gaza* (London; New York: Verso, 2012); Irm Haleem, *The Essence of Islamist Extremism: Recognition through Violence, Freedom through Death* (London; New York: Routledge, 2011); Adonis, *Violence and Islam: Conversations with Houria Abdelouahed* (Malden, MA: Polity, 2016); Asma Afsaruddin, *Striving in the Path of God: Jihad and Martyrdom in Islamic Thought* (Oxford, New York: Oxford University Press, 2013).

26 Talal Asad, *On Suicide Bombing* (New York: Columbia University Press, 2007), 4.

27 Bart Schuurman, Peter Grol, and Scott Flower, *Converts and Islamist Terrorism: An Introduction*, ICCT Policy Brief (The Hague: International Centre for Counter-Terrorism, June 2016).

28 I develop this thought in an article entitled "On violence in Islam" published in Polish in Znak monthly. Patrycja Sasnal, "O przemocy w Islamie," *Znak*, nr 741, February 2017, accessed March 28, 2019, www.miesiecznik.znak.com.pl/o-przemocy-w-islamie/.

29 Maxime Rodinson, *Europe and the Mystique of Islam* (London; New York: I.B. Tauris, 2006), xiii–xiv.

30 Part of the findings included in Chapter 4 were first published in a EuroMeSCo Joint Policy Study, issued in the framework of the project "Euro-Mediterranean Political Research and Dialogue for Inclusive Policymaking Processes and Dissemination through Network Participation," co-funded by the European Union and the European Institute of the Mediterranean. See Patrycja Sasnal, "The Reasons for Radical Groups' Appeal among European and Arab Citizens: The Case of ISIS," in *Terrorist Threat in the Euro-Mediterranean*, ed. Amal Mukhtar (Barcelona: IEMed, 2016) 7–21.

Bibliography

Adonis. *Violence and Islam: Conversations with Houria Abdelouahed*. Malden, MA: Polity, 2016.

"Afghan Civilians | Costs of War." Watson Institute of International and Public Affairs, Brown University. Accessed December 1, 2017. http://watson.brown.edu/costsofwar/costs/human/civilians/afghan.

Afsaruddin, Asma. *Striving in the Path of God: Jihad and Martyrdom in Islamic Thought*. Oxford, New York: Oxford University Press, 2013.

Arendt, Hannah. *On Violence*. Orlando; Austin; New York; San Diego; London: HMH, 1970.

Asad, Talal. *On Suicide Bombing*. New York: Columbia University Press, 2007.

Ayyash, Mark Muhannad. "The Paradox of Political Violence." *European Journal of Social Theory* 16, no. 3 (August 1, 2013): 342–56.

Bar On, Bat-Ami. *The Subject of Violence: Arendtean Exercises in Understanding*. Lanham, MD: Rowman & Littlefield, 2002.

Bernstein, Richard J. *Violence: Thinking without Banisters*. Cambridge, UK: Polity, 2013.

Bloxham, Donald, and Robert Gerwarth, eds. *Political Violence in Twentieth-Century Europe*. Cambridge: Cambridge University Press, 2011.

Brachman, Jarret M. *Global Jihadism: Theory and Practice*. London; New York: Routledge, 2008.

Calhoun, Craig J., and John McGowan. *Hannah Arendt and the Meaning of Politics*. Minneapolis: University of Minnesota Press, 1997.

Curthoys, Ned. "The Refractory Legacy of Algerian Decolonization: Revisiting Arendt on Violence." In *Hannah Arendt and the Uses of History. Imperialism, Nation, Race, and Genocide*, edited by Richard King and Dan Stone, 109–29. New York, Oxford: Berghahn Books, 2007.

Fanon, Frantz. *The Wretched of the Earth*. New York: Grove Press, 2007.

Finlay, Christopher J. "Hannah Arendt's Critique of Violence." *Thesis Eleven* 97, no. 1 (May 2009): 26–45.

Frazer, Elizabeth, and Kimberly Hutchings. "On Politics and Violence: Arendt Contra Fanon." *Contemporary Political Theory* 7, no. 1 (February 1, 2008): 90–108.

Gibson, Nigel. "Is Fanon Relevant? Toward an Alternative Foreword to 'The Damned of the Earth.'" *Human Architecture: Journal of the Sociology of Self-Knowledge* 5, no. 3 (June 21, 2007). https://scholarworks.umb.edu/humanarchitecture/vol5/iss3/6.

Gines, Kathryn T. "Arendt's Violence/Power Distinction and Sartre's Violence/Counter-Violence Distinction: The Phenomenology of Violence in Colonial and Post-Colonial Context." In *Phenomenologies of Violence*, edited by Michael Staudigl, 123–44. Leiden: Brill, 2013.

———. *Hannah Arendt and the Negro Question*. Bloomington: Indiana University Press, 2014.

Gray, John. "Steven Pinker Is Wrong about Violence and War." *The Guardian*, March 13, 2015, www.theguardian.com/books/2015/mar/13/john-gray-steven-pinker-wrong-violence-war-declining.
Haleem, Irm. *The Essence of Islamist Extremism: Recognition through Violence, Freedom through Death*. London; New York: Routledge, 2011.
Herzog, Annabel. "Hannah Arendt's Concept of Responsibility." *Studies in Social and Political Thought* 10 (August 2004): 39–56.
———. "The Concept of Violence in the Work of Hannah Arendt." *Continental Philosophy Review* 50, no. 2 (2017): 165–79.
Hudis, Peter. *Frantz Fanon: Philosopher of the Barricades*. London: Pluto Press, 2015.
Huntington, Samuel P. *The Third Wave: Democratization in the Late Twentieth Century*. Norman: University of Oklahoma Press, 1993.
"Hypocrisy | Definition of Hypocrisy in English by Oxford Dictionaries." *Oxford Dictionaries | English*, https://en.oxforddictionaries.com/definition/hypocrisy.
Lee, Christopher J. *Frantz Fanon: Toward a Revolutionary Humanism*. Athens: Ohio University Press, 2015.
Owens, Patricia. *Between War and Politics: International Relations and the Thought of Hannah Arendt*. Oxford: Oxford University Press, 2007.
Parsons, Tim, and JH Bloomberg School of Public Health. "Updated Iraq Survey Affirms Earlier Mortality Estimates." Johns Hopkins Bloomberg School of Public Health. Accessed December 1, 2017. www.jhsph.edu/news/news-releases/2006/burnham-iraq-2006.html.
Pinker, Steven. *The Better Angels of Our Nature: Why Violence Has Declined*. New York: Viking, 2011.
Rodinson, Maxime. *Europe and the Mystique of Islam*. Translated by Roger Veinus. London; New York: I.B. Tauris, 2006.
Sasnal, Patrycja. "O przemocy w islamie," *Znak*, nr 741, February, 2017. Accessed March 28, 2019. www.miesiecznik.znak.com.pl/o-przemocy-w-islamie/.
———. "The Reasons for Radical Groups' Appeal among European and Arab Citizens: The Case of ISIS," In *Terrorist Threat in the Euro-Mediterranean*, edited by Amal Mukhtar, 7–21. Barcelona: IEMed, 2016.
Schuurman, Bart, Peter Grol, and Scott Flower. *Converts and Islamist Terrorism: An Introduction*. ICCT Policy Brief. The Hague: International Centre for Counter-Terrorism, June 2016.
Swift, Simon. *Hannah Arendt*. London; New York: Routledge, 2009.
———. "Hannah Arendt, Violence and Vitality." *European Journal of Social Theory* 16, no. 3 (August 2013): 357–76.
Weizman, Eyal. *The Least of All Possible Evils: Humanitarian Violence from Arendt to Gaza*. London; New York: Verso, 2012.
"Yemen Emergency." *World Food Programme*. Accessed March 21, 2019. www1.wfp.org/emergencies/yemen-emergency.
Zeilig, Leo. *Frantz Fanon: The Militant Philosopher of Third World Revolution*. London; New York: I.B. Tauris, 2016.

1 Fanon and Arendt

A black man and a Jewish woman

They seemed to differ in almost everything. Frantz Fanon, 19 years younger than Arendt, was a black Martinican-French psychiatrist (later Algerian by choice) who devoted his adult life to treating patients, analysing racism and helping the Algerian independence cause, which he made into a general Third World independence from the colonist. Arendt was a Jewish-German philosopher who spent her life teaching students and analysing the human condition in the West, later in her years in the United States. Two messiahs of two different worlds: the dispossessed black and brown Third World and the dominant white Western world. The uniqueness of Fanonian experience is perhaps marked the most by his blackness. The strength of the skin colour as a social trait does not compare with any other, not even with underprivileged gender. As influential public figures at the time Fanon and Arendt have become to be easily categorised, their philosophies – simplified.

The two had much more in common than reveals itself at first sight. Both were escapees from the milieus that betrayed them: as a black man Fanon escaped racist and colonist France, as a Jew Arendt escaped Nazi Germany. Both were fighters in their own right: Arendt in the US fighting with her pen, Fanon in Algeria fighting with his pen and skills as a medical doctor. Both experienced the horrors of the Second World War: Fanon in the French army (he went there to fight fascism for France and ended up fighting racism in France), Arendt in the Gurs internment camp. Both hailed from the oppressors and the oppressed at the same time. Fanon was French and yet he was black – Arendt was German and yet she was a Jew (and a woman). Both ran away from the oppressor part of their identity. Fanon quit his French identity and became Algerian – changing his name to Omar Fanon in 1958, Arendt abandoned her German identity for the American one. Both loved their mother tongue and revelled in immersing themselves in French and German, respectively – the languages that housed the origins of racism and fascism. Both continue to stir interest and the proliferation of books about their lives, activism and philosophies is unyielding.

Fanon writes WotE on his death bed. In fever and anxiety. Parts of the book are made of his former statements at the African summits. The original work is the chapter about violence. He has done no research except for his lifelong readings and experience. What is telling for the way WotE sounds is Claude Lanzmann's recollection of his first meeting with Fanon: "death (…) gave his every word the power both of prophecy and of the last words of a dying man."[1] WotE is a deeply personal work, a last gasp uttered as loudly as possible. Writing might have meant the same for Arendt – EiJ was an "intensely personal work,"[2] she admitted. OV is short and not particularly novel – it strikes the tones already reverberating in OoT, HC, EiJ, OR and *Crises of the Republic.*

The book was inspired by events, perhaps by fear. Arendt admits that she writes it in the context of the 20th century violence (OV, 3). She witnesses it in the US where daily violence occurs at the universities and the cold war rages, although she does not write it to students. Arendt is already an established, widely read academic who can influence opinions. In OV she also, or perhaps primarily, takes an overdue position in the debate that unfolded among French intellectuals in the 1950s about the dilemma that leftist thinkers found in the terror of the oppressed in Algeria.[3] Fanon writes WotE between Algeria and Tunisia, where he treated patients at psychiatric hospitals in Blida and then Tunis. He is engulfed by violence not least in his medical milieu but also in his contacts with the Front de la Libération Nationale (FLN) as their spokesperson.

Both develop a unique style of writing – distinct at first sight but not without similarities. Fanonian language is poetic, lofty and religious, almost sacral but meaty rather than ephemeral ("the almighty body of violence," "the last shall be first," "Europe's tower of opulence" – WotE, 50, 2, 57). One explanation for it is that Fanon, himself a messiah of the dispossessed, proposes a religious brotherhood, a mystical doctrine.[4] He describes the coloniser in biological and animalistic terms ("parasites," WotE, 7). At times his language turns mystical, as if taken from fables (opulence, diamonds, silk, cotton, oil, gold: WotE, 58). He is partially a fiction writer and a poet. When asked by one of his translators to explain a passage he responded: "This passage is inexplicable. When I write such things I seek to touch my reader in his emotions, i.e. irrationally, almost sensually."[5] The visceral naturality of Fanon's language can easily impress on the reception of the text. The feverish tone corresponds not only with the state of the author's body at the time but also with the boldness of his project. His message was to inspire the creation of a new man – in what other language can this endeavour be given an initial push if not that of prophets, poets and philosophers?

WotE's style seems to contrast with the calm, reflective tone of Arendt's works, including OV. If Fanon is interested in the sound of words, Arendt seems to be preoccupied with their meaning. But Arendt's tranquillity is illusive. It's marked by subdermal fear and an equally daring project that she had embarked on. Like Fanon's, her language is literary, even if inspired by different literature than Fanon's. In the mottos and in-text literary insertions she makes her drama unfold effortlessly but surely. Fanon also avails himself with literary quotations (frequently from Aimé Césaire[6]) but he equally employs the style into his own writing. One example of this difference in technique but similarity in tone is the comparison between Arendt's opening of her book about the trial of Eichmann with a quotation from Bertold Brecht's: "O Germany, …, but whoever sees you, reaches for his knife"[7] and an almost analogical in tone and meaning quotation of Fanon's words from WotE: "When the colonized hear a speech about Western culture they draw their machetes or at least check to see they are close at hand."[8]

Philosophical influences: philosophers of beginnings

When studying medicine in Lyon (1947–51) Fanon read Sigmund Freud, Carl Gustav Jung, Jacques Lacan, but also Immanuel Kant, Søren Kierkegaard, Karl Jaspers, Claude Lévi-Strauss, Georg Wilhelm Friedrich Hegel and the young Karl Marx[9] – we also know that a couple of years later in Algeria he was given Friedrich Engels's *Anti-Duhring*. In 1947/8 he attended lectures by Maurice Merleau-Ponty at the University of Lyon. Merleau-Ponty lectured then on the unity of soul and body in the philosophies of Nicolas Malebranche, Maine de Biran and Henri Bergson.[10] Fanon never talked to him in person, but he read the journals that Jean Paul Sartre established with Merleau-Ponty: *Les Temps Modernes* and *Présence Africaine*. There is a consensus that Sartre imprinted the strongest influence on Fanon[11] and so the role of Maurice Merleau-Ponty is usually, with minor exceptions,[12] downgraded. However, phenomenology, rather than existentialism, suited his experience as a black man and his thoughts on racism. In *Phenomenology of Perception* Merleau-Ponty develops a conception of consciousness that comes from the bodily interaction with the world, in which a body is a form of consciousness.[13] The book, published the moment Fanon found himself in Lyon attending Merleau-Ponty's lectures, influences Fanonian thought in the prominence that it gives to the body and experience. In Fanon's writing on racism reverberate the words of Merleau-Ponty: "the subject has simply the external world that he gives himself."[14] Existence, meaning being in the world through a body, serves as a basis for Fanon's understanding of human experience. In *Black Skin, White Masks* (BSWM), published nine years before WotE and before Fanon's Algerian adventure, the central concept is the lived experience, *expérience vécue*, used by Merleau-Ponty and common in psychiatric research at the time. Fanon cites Merleau-Ponty to say that the black man does not see himself "normally" – he abnormalises himself, while the white man is both a mystifier and mystified.[15] What remains in WotE of Merleau-Ponty's thought is the accent on the body and experience that allows the man of colour to be free and give himself the external world.

> The density of History determines none of my acts. I am my own foundation. And it is by going beyond the historical and instrumental given that I initiate my cycle of freedom.
>
> (BSWM, 204–5)

In WotE, indeed, it is Sartre's *Critique de la raison dialectique* that imprints its mark most ostentatiously. Fanon read the book the moment it was published in 1960 and is known to have later given lectures about it to Algerian forces.[16] In 1961 he asked his editor – François Maspero – that he convinces Sartre to write the preface to WotE. The two men, and Simone de Beauvoir, had a couple of intense meetings in Rome (facilitated by Lanzmann), where, on the first occasion, they would talk to early morning hours or, another time, see Fanon dying. Fanon seemed to disprove of the French couple's lifestyle – getting up late, enjoying meals and drinks – which contrasted with his own ascetism. De Beauvoir found him full of life with personal horror of violence and counterviolence – Fanon thought this was his failing, his weakness as an intellectual.[17] Contrary to his popular image, he did not pick up a gun since the Second World War.

Throughout his life Fanon, as a psychiatrist, was also under a strong influence of psychoanalysis but in WotE he finally and clearly gives precedence to social and political considerations over an individual psyche. In WotE, in particular, Fanon is preoccupied with the unity of the revolutionary group, with solidarity or lack thereof. Sartre's

Critique dwells on that point of utmost interest to Fanon at the time: how revolutionary group cohesion crumbles. It also validates violence: "it suddenly reactualises violence as the intelligible transcendence of individual alienation by common freedom."[18]

Fanon borrows from Sartre the dialectical couple of the colonist and the colonised, although it was the Hegelian master and slave allegory that initially shaped it:[19]

> the colonialist discovers in the native not only the Other-than-man but also his own sworn Enemy (in other words, the Enemy of Man). This discovery does not presuppose resistance (open or clandestine), or riots, or threats of revolt: the violence of the colonialists itself emerges as an indefinite necessity or, to put it another way, the colonialist reveals the violence of the native, even in his passivity, as the obvious· consequence of his own violence and as its sole justification (…) the colonialist and the native are a couple, produced by an antagonistic situation and by one another.[20]

On the basis on this figure Fanon will later justify the struggle of the colonised, which he will call the "absolute praxis,"[21] taking directly from Sartre's *Critique* again:

> important for us is to show that dialectic, as the controlled development of praxis, cannot experience itself (either as constituent or as constituted) except in and through the praxis of struggle, that is to say, antagonistic reciprocity.[22]

Finally, Fanon's gist idea of the Third World man creating the future new humanity through his struggle can also be traced back to Sartre:

> The 'wretched of the earth' are precisely the only people capable of changing life, and who do change it every day, who feed, clothe and house humanity as a whole.[23]

The influence is not necessarily one way. Fanon, fascinated with Sartre and a devoted reader of most of his texts is sometimes credited with influencing Sartre in seeing a system in racism. Bernasconi suggests that it was first Fanon who developed Sartre's racism as "gaze" in his earlier writings into racism as a system, which can be seen both in *Critique de la raison dialectique* and WotE.[24]

Fanon's philosophy rests on responsibility – personal responsibility for the humanity, its marvels and crimes alike. "Peloponesian War is as much mine as the invention of the compass" (BSWM, 200). Despite the experience of otherness and discrimination Merleau-Ponty's universalism reverberates in Fanon's words. We are born into a world of otherness: one's own subjectivity raises from this initial unity into which we come to exist.

> In the cultural object, I feel the close presence of others beneath a veil of anonymity. … In reality, the other is not shut up inside my perspective of the world, because this perspective itself has no definite limits, because it slips spontaneously into the other's, and because both are brought together in the one single world in which we all participate as anonymous subjects of perception.[25]

The other skin colour is an element of the subject of perception, yet in the cultural objects – the printing press, guns, books – the colourfulness disappears. How is it possible, Fanon seems to be asking, that cultural objects are perceived white?

It has been pointed out that Sartre's existential ethics impacted Fanon, specifically in ideas about a nation-state that respects, and more still, is constituted on difference.[26] The roots of Fanonian concepts of a nation and his philosophy of commitment (on the 1 page of BSWM he imprecisely quotes Marx's 11th thesis on Feuerbach) stem from interest in equality and discrimination and, again, go back to Sartre and his *Anti-Semite and Jew*. Fanon's teacher in Martinique instructed him to be alert whenever he comes across anti-Semitism because "the Jew was him."[27] The Jew – "my brother in misfortune", Fanon would say:

> just as the Jew who is lavish with his money is suspect, so the black man who quotes Montesquieu must be watched. (…) Colonial racism is no different from other racisms. Anti-Semitism cuts me to the quick; I get upset; a frightful rage makes me anemic; they are denying me the right to be a man. I cannot dissociate myself from the fate reserved for my brother.
>
> (BSWM, 68–70)

Still the Jew is white and, contrary to him who may be slave of the idea that others have of him, Fanon is the slave of his appearance. Fanon has to be considered one of the first to take into account the ethnic-racial relations in philosophy.

It may come as a surprise, however, that Fanon was able to see the commonality between various forms of discrimination but failed to see it vis-à-vis women. He had to be convinced by his friends to recognise his daughter from an affair that he had with Michelle B. in Lyon and would later see the woman's place at home with the children. Even though it was his wife, Josie, who typed his texts (himself he never learnt to type).[28]

Arendt's philosophical upbringing is inseparable from the Second World War. She is "the theorist of beginnings" as Margaret Canovan calls her in the introduction to the Human Condition,[29] a book written a decade earlier than OV. A new beginning is what Fanon will be interested in too. Arendt captures humanity in a still at the beginning of great actions that man is capable of undertaking in concert. Fanon would want one of those stills be taken in Algeria of 1960. What differentiates the two philosophers is that Arendt has a life vest sewn from two elements: Martin Heidegger's deconstructive hermeneutics and Walter Benjamin's historiography. The life vest is the ancient philosophical tradition, by rediscovering which she rebuilds thought after the rupture of totalitarianisms. She recovers the original meaning, obscured by tradition, in Western philosophical thought from the Greeks. One philosophical influence that is of utmost importance to Arendt and to this study on political violence is the Aristotelian *bios politikos*, "the realm of human affairs, stressing the action, *praxis*, needed to establish and sustain it. Neither labor nor work was considered to possess sufficient dignity to constitute a *bios* at all, an autonomous and authentically human way of life." (HC, 13). Arendt constructs her *vita activa* and, eventually, her understanding of public affairs on the basis of this Aristotelian concept.

On violence specifically, Arendt should be seen as engaging in intellectual debates not only with Fanon, Sorel and Pareto – as she explicitly does in OV. Her take on violence, most probably, was also informed by Walter Benjamin and Albert Camus. She conspicuously omits any mention of Benjamin's *Critique of Violence* in OV, even though she was most probably well familiar with it.[30] If judged by the reading of OV only, Arendt does not share Benjamin's understanding of violence, which has two functions: a law-making and law-preserving one.[31] To dissolve "the link between violence and the law" – as Giorgio Agamben put it[32] – Benjamin introduces "divine violence" – "revolutionary deactivation as pure immediate violence that deposes the law"[33] – which breaks the cycle between violence and law, and abolishes state power.[34] If Arendt's take on violence is expanded to OR, as Finlay has done, then both Arendt and Benjamin can be in agreement in wanting to break the law-violence linkage.[35] Isaac (and Carroll) argue that on the question of violence Arendt sides with Camus: they shared a pessimistic although courageous attitude to the tragedies on the 21st century and both advocated a self-restricting democratic empowerment.[36]

Fanon does not have a similar life vest because he cannot. Above all ancient philosophy is a white man's philosophy, not the slave's. And since man is his own creation, he does not want to have it:

> We would be overjoyed to learn of the existence of a correspondence between some black philosopher and Plato. But we can absolutely not see how this fact would change the lives of eight-year-old kids working in the cane fields of Martinique or Guadeloupe.
>
> (BSWM, 204)

His idea is about a fresh start and a new world. This thought has an important function: the lifting of the spirit. Arendt uplifts too – her philosophy slowly lifts the man from the rubble of the Second World War and allows him to see another human being as a companion in action. Together, as new citizens, they will start political life all over. In that Arendt was closer to the modernist Islamic thinking about the community and violence than Fanon. Although they come from almost incomparable milieus and traditions, unwittingly she echoes the works of the shaykh of al-Azhar Muhammad Abduh, considered the main Muslim reformer, whose sole objective was "elevating the Muslim community (*umma*) to a higher level, strengthening its ethics, awakening it socially, gradually and with patience, without violence."[37]

Like Fanon Arendt is difficult to classify in any current of socio-political thought. She was clearly inspired by Heidegger's and Edmund Husserl's lectures, her cooperation with Karl Jaspers and the rereading of the ancient Greeks but she follows no single tradition of thought. She is seen as a thinker on political nonviolence. The reasons for that can be found further in this paper, but elsewhere she also praises the experience of war, in which a man is "fully alive and political."[38] These are not Arendt's words but Owens's, who reads this message in at least three works of Arendt's: *On Revolution*, *Between Past and Future* and in an introduction to Glenn Gray's *The Warriors*. The thought is perhaps disputable in the definite opinion presented by Owens but indeed the attractiveness of the experience of war for many people comes from the self-disclosure in violent action or self-display, Arendt would claim, which is to live.[39]

In OV Arendt objects to Fanon's embracement of violence. She sees George Sorel as Fanon's philosophical master and connects his theory of violence to Henri Bergson's *élan vital* or *élan original*: "an internal push that has carried life, by more and more complex forms, to higher and higher destinies."[40] To Sorel violence was that impetus that gives class struggle its energy.[41] Indeed Fanon most probably read Sorel's *Reflections on Violence* but the connections are thin, perhaps with the exception of the views on the bourgeois society (discussed below) and common roots (Hegelianism and the absoluteness of class or race identity,[42] although these identities can by no means be commensurate). Sorel's Marxism and syndicalism could be useful for Fanon if the fight for Algerian independence was a class struggle. The ethnic relations in Algeria were, however, so particular and the absence of proletariat so obvious that orthodox Marxism proved of lesser use. Which is not to say that Fanon abandoned dialectics – as Sekyi-Otu eloquently laid out, a dialectical reading of Fanon reveals a very coherent thinker on dialectics of experience.[43]

In passing and indirectly several authors have already noticed the similarities between Fanon and Arendt. Cocks thinks that Arendt's OoT anticipated WotE[44] because of the observation in OoT that the metropolis had to apply violence as a last resort to remain in control. Also, in OoT Arendt is clearly preoccupied with colonialism and imperialism, seeing in them the signs of totalitarianism.[45] In OoT Arendt makes the case that the advancement of totalitarianisms was, i.a. "the boomerang effect of imperialism upon the homeland" (OoT, 155). In another article Cocks compared the anti-nationalistic stances of Arendt, Fanon and Rosa Luxemburg. Recently it has been suggested that at least Arendt had more in common with Fanon than Schmitt.[46]

Reception of WotE and connection with jihadism

The reception of Fanon was immediately articulate. Jean Daniel, the founder of *Nouvel Observateur*, France's well known weekly, was particularly critical of Sartre's preface and his "masochistic fury." He called WotE "a terrible portent of barbaric dispensers of justice," adding prophetically that "disciples of these theses will be tranquil killers, self-justified executioners, terrorists."[47] Indeed, people like Sartre would support the Maoist groups that hailed from North Africa and terrorised France in 1970s.[48] If reaction on the Left was diverse it was uniquely hostile on the Right: Gilbert Comte, a writer for *La Nation française*, called it the "Mein Kampf of decolonization."[49] Even Jacques Azoulay, who admired and befriended Fanon, distanced himself from his master having read WotE.[50]

With or without Sartre's preface Fanon's rhetoric fitted in the cultural upbringing of activists and fighters who were in 1960s and 1970s in one way or another struggling with "the oppressor," while giving "the oppressed" a sense of agency. Scholars tend to refer to WotE as a "formative text" for revolutionaries.[51] However, the level and kind of influence of Fanon's WotE on them was different. Ali Shariati in Iran and Steven Biko in South Africa adjusted Fanon's theories to their own needs and circumstances. He also impacted the Black Panthers in the US and, to a much lesser extent the IRA in Ireland.[52] It would be difficult, however, to prove a direct intellectual linkage with Fanon of many revolutionary, liberation or terrorist movements and organisations.

Outside of Algeria Fanon was first and best known in the Muslim world in Iran thanks to Ali Shariati, an Iranian intellectual who studied in Paris, feverishly read Sartre and was in contact with Fanon. He later translated BSWM and WotE to Farsi. Shariati adopted Fanon's struggle of the colonised to the dichotomy in Islam between the arrogant and the weak, which would later be used by Ruhollah Khomeini in the Islamic revolution of 1979. Shariati, under Heidegger's influence, thought a relationship with the religious past was a necessary element of the creation of the new man. Fanon disregarded the past, perhaps specifically the religious past.[53] He neither advocated sacrifice as martyrdom: the revolutionary should live and act! Shariati would rather see martyrdom as an example and guidance for the masses.[54] In a letter to Shariati Fanon explicitly voiced his reservations about stirring religious fervour:

> Even if I do not share the same sentiments with regard to Islam as you, I respect your idea which affirms that in the Third World (I prefer to say in the Near and Middle East, if you allow me to) Islam, more than other social and ideological forces, had anticolonial potential and anti-Western nature. I hope that you intellectuals can give life to the inert and drugged body of the Islamic Orient so that its people are awake to create another man and another civilization. For my part, I worry that **the revitalized sectarian and religious spirit will hold off a self-forming nation from its future and petrify it on its past**.[55]

Religions are exclusivist which went against Fanonian respect for difference and secularism. But Fanon's work would live its own life in Iran once Shariati translated WotE. A sentence from Fanon was on Iranian banners in 1979: "The chador is a thorn in the eye of Western imperialism," legitimising the opposite of what Fanon called for, two years after Shariati's death.

The Iranian revolution was a revolution of the Shia Muslims, against whom extremist Sunnis would pick up a fight. Yet it has been suggested that the tones of Shariati, specifically his term of "westoxication," when he described the condition of the Westernised Iranian elite, would be later borrowed by Sayyid Qutb,[56] considered the father of jihadism, a phenomenon in Sunni Islam. Qutb called the Egyptians that abandoned their Egyptian identity and became Westernised "the brown English." Could have Fanon influenced modern jihadism this way?

Frantz Fanon's grounding in the Arab and Islamic culture was not profound. He did not speak Arabic. When he moved to Tunisia from Algeria in 1957 he continued working as a doctor and began writing for the FLN's paper El Moudjahid, although he had little real contact with the organisation.[57] The bulk of his time was not devoted to freedom fighting but treating patients at the hospitals first in Blida (Algeria) then in Tunis, which he reformed and reorganised. Fanon contributed to El Moudjahid – the title meant a fighter in a jihad – but instead of waging a "holy war" the editors of the paper explained that jihad meant a "dynamic manifestation of self-defense designed to preserve or recover a heritage of higher values that are indispensable to both individual and community. *Jihad* is the quintessence of liberal and open patriotism."[58] It suited Fanon's understanding of the struggle but in reality the jihad was often understood differently, as Muslims fighting Christians.

While Fanon was active in Algeria and Tunisia up until his death a new post-colonial world was already shaping up elsewhere in the Middle East. In the nearby Egypt, in 1952 Gamal Abdel Nasser took power in a military coup. Initially he sided with the Islamists from the Muslim Brotherhood (MB) but, afraid of the competition, he immediately cracked down on them once in power. Sayyid Qutb, a MB member, was in 1954 convicted of plotting to assassinate Nasser. With short intervals he spent a decade in prison, where he was tortured and, eventually, hanged in 1966. Qutb would later become the founding father of jihadism. He wrote *Milestones* (*Ma'alim fit-tariq*), which influenced Sunni jihadists, including the current head of Al-Qaeda Ayman az-Zawahiri who joined the MB shortly before Qutb's execution.

Even though there is no evidence that Qutb read Fanon at all, several authors go as far as to make the connection between them.[59] Because of the general similarities in the revolutionary philosophy of the slave and master others explicitly extrapolate Fanon's influence over contemporary terrorist organisations, such as Al-Qaeda.[60] Qutb's vision does have messianic undertones but his remedy to the diagnosis of a sick society, similar to the diagnosis of Fanon, is diametrically different from Fanon's. God is Qutb's saviour, not human agency:

> The leadership of mankind by Western man is now on the decline ... because it is deprived of those life-giving values which enabled it to be the leader of mankind ... Islam is the only system which possesses these values... Only in the Islamic way of life do all man become free from the servitude of some man to others and devote themselves to the worship of God alone, deriving guidance from Him alone and bowing before Him alone.[61]

In *Hatha ad-Din* Qutb said that a Muslim, by struggling against others, is also struggling against himself:

> horizons are opened to him in the faith which would never be opened to him if he were to sit immobile and at rest... His soul, feelings, his imagination, his habits, his nature, his reactions and responses – all are brought to a point of development which he could not have attained without hard and bitter experience.[62]

Such self-struggle does bring to mind the dialectical praxis of the colonised, but for Fanon the ultimate goal is freedom, for Qutb – puritanism, first of all, and only through it – freedom.[63] In a recent work on the ideological struggle between Nasser and Qutb, Gergis finds direct linkages between Fanon and Nasser, not with Qutb; Nasser's *The Philosophy of the Revolution* shares Fanonian criticism of the elite and educated classes, the role of agency and culture in emancipation and the destructive function of social subordination.[64]

Today Fanon is read at the universities and in English, rarely in Arabic at Egyptian or Algerian universities. Apparently, there is more research about the university courses and readership of Arendt,[65] and certainly Sartre's popularity in the Arab world, at least until 1967, cannot be matched.[66] But Fanon is still little known in the Arabic-speaking world, both in academia and in political circles, even if WotE was already translated into Arabic in 1963 and published in Lebanon. It has been suggested that the demand for a translation might have come from the Palestinians in Lebanon who were at the forefront of the Arab-Israeli struggle.[67] Algeria was to become a model to follow for Yaser Arafat and the Palestinian national struggle,[68] which does not necessarily corroborate the thesis that Fanon's writings influenced the Palestinian leadership.[69]

WotE's greatest fame boomed in the United States. The book was translated twice, in 1963 and 1965, and was immediately noted on the American market. The position and role of violence stirred a lesser debate there than in France. WotE became the "bible" of the black Americans, having profound impact on two black movements: the Black Panthers and Black Power. The influential Malcolm X, a member of the Nation of Islam, with which X got disillusioned, is known to have visited Fanon when still in Algeria and when Malcom X was still Robert Williams.[70] For the black Americans, who felt colonised by the whites and wanted to find dignity in the struggle for their rights, who felt that the division is not as black and white as it might seem and one needs to be wary of some of the black elite, Fanon's work was a gift. On the level of praxis everything clicked here.

It was because of the book's popularity in the black movements that Arendt read Fanon and related his thoughts to the violence that occurred on campuses in the US in the 60s. Since then and because so many have already seen Fanon as influence on revolutionaries and terrorists alike, it is also easy – perhaps too easy – to relate him to the jihadist violence in the 21st century.

Until Macey's book of 2000 Alice Cherki's portrait was considered one of the best if not the best biography of Fanon. Cherki does not even mention Arendt – her OV and polemics with him were not broadly known and not as influential a work as Fanon's translations in the US. Arendt's endeavour in OV is to debate violence primarily not in relation to Fanon, although he is mentioned, but as political violence in its generality. Arendt takes Fanon for what he proposes in WotE only, but she takes him as a philosopher.

Notes

1 Claude Lanzmann, *The Patagonian Hare: A Memoir*, trans. Frank Wynne (New York: Farrar, Straus and Giroux, 2013), 338. Lanzmann tells of a mesmerizing encounter with an ascetic "visionary" in an apartment with bare walls and no furniture.

2 Amos Elon, "Introduction," in *Eichmann in Jerusalem. A Report on the Banality of Evil*, Hannah Arendt (New York: Penguin, 2006), IX.

3 Ned Curthoys, "The Refractory Legacy of Algerian Decolonization: Revisiting Arendt on Violence," in *Hannah Arendt and the Uses of History. Imperialism, Nation, Race, and Genocide*, ed. Richard King and Dan Stone (New York; Oxford: Berghahn Books, 2007), 109–29.

4 Homi K. Bhabha, "Framing Fanon," in *The Wretched of the Earth*, by Frantz Fanon (New York: Grove Press, 2007), IX.

5 Richard Philcox, "On Retranslating Fanon, Retrieving a Lost Voice," in *The Wretched of the Earth*, by Frantz Fanon (New York: Grove Press, 2007), 242.
6 Fanon's intellectual father in Martinique and author of *Discourse on Colonialism*, a formative text for négritude literary movement. Aimé Césaire, *Discourse on Colonialism*, trans. Joan Pinkham (New York: Monthly Review Press, 2001).
7 Hannah Arendt, *Eichmann in Jerusalem* (New York: Penguin, 2006), 2.
8 Frantz Fanon, *The Wretched of the Earth* (New York: Grove Press, 2007), 8.
9 Peter Hudis, *Frantz Fanon: Philosopher of the Barricades* (London: Pluto Press, 2015), 21.
10 Jacek Migasiński, *Merleau-Ponty*, Myśli i Ludzie (Warszawa: Wiedza Powszechna, 1995), 25.
11 David Caute, *Frantz Fanon* (Vichy: Seghers, 1970), 140.
12 For an exemplary exception see: Caute, 140.
13 Komarine Romdenh-Romluc, *Routledge Philosophy GuideBook to Merleau-Ponty and Phenomenology of Perception* (London; New York: Routledge, 2010), 2–3.
14 Maurice Merleau-Ponty, *Phenomenology of Perception* (London: Psychology Press, 2002), 507. In the better, Polish translation it reads: "Podmiot ma tylko takie zewnętrze, jakie sam sobie nada" [The subject has only such outside as it gives itself]. See Maurice Merleau-Ponty, *Fenomenologia Percepcji*, trans. Małgorzata Kowalska and Jacek Migasiński (Warszawa: Fundacja Aletheia, 2001), 457.
15 Frantz Fanon, *Black Skin, White Masks*, trans. Richard Philcox (New York: Grove Press, 2008), 200.
16 David Macey, *Frantz Fanon: A Biography* (London; New York: Verso, 2012), 448–49.
17 Macey, *Frantz Fanon*, 453–57.
18 Jean Paul Sartre, *Critique of Dialectical Reason*, vol. 1 (London; New York: Verso, 2004), 433.
19 See chapter "The Black Man and Hegel" in BSWM, 191–97.
20 Sartre, *CDR*, 1:720–21.
21 Fanon, *WotE*, 43.
22 Sartre, *CDR*, 1:804.
23 Sartre, 1:241.
24 Robert Bernasconi, "Fanon's 'The Wretched of the Earth' as the Fulfillment of Sartre's 'Critique of Dialectical Reason'," *Sartre Studies International*, no. 2 (2010): 38–39.
25 Merleau-Ponty, *Phenomenology of Perception*, 405, 411.
26 Pramod K. Nayar, *Frantz Fanon* (London; New York: Routledge, 2012), 33–36.
27 Macey, *Frantz Fanon*, 2012, 166. For an analysis of Fanon, Arendt, postcolonialism and anti-Semitism see Bryan Cheyette, "Postcolonialism and the Study of Anti-Semitism," *American Historical Review* 123, no. 4 (October 2018): 1234–45.
28 See Leo Zeilig, *Frantz Fanon: The Militant Philosopher of Third World Revolution* (London; New York: I.B. Tauris, 2016), 32–33; David Macey, *Frantz Fanon: A Biography* (London; New York: Verso Books, 2012).
29 Hannah Arendt, *The Human Condition* (Chicago, IL; London: The University of Chicago Press, 1998), VII.

30 Even though none of this information proves her knowledge of Benjamin's text it was already published in 1921, Arendt knew him personally and shortly before her death prepared a volume of his works, including "Critique of Violence," for publication.

31 Walter Benjamin, *Reflections: Essays, Aphorisms, Autobiographical Writings*, ed. Peter Demetz (New York: Schocken, 1986), 284.

32 Giorgio Agamben, *Homo Sacer. Sovereign Power and Bare Life* (Redwood City, CA: Stanford University Press, 1998), 42.

33 Sami Khatib, "Towards a Politics of 'Pure Means': Walter Benjamin and the Question of Violence," *Anthropological Materialism* (blog), August 28, 2011, accessed March 28, 2019, https://anthropologicalmaterialism.hypotheses.org/1040.

34 Benjamin, *Reflections*, 300.

35 Christopher J. Finlay, "Arendt's Critique of Violence," *Thesis Eleven* 97, no. 1 (May 2009): 26–45.

36 Jeffrey C. Isaac, *Arendt, Camus, and Modern Rebellion* (New Haven, CT: Yale University Press, 1992); David Carroll, *Albert Camus the Algerian: Colonialism, Terrorism, Justice* (New York: Columbia University Press, 2007).

37 'Uthman Amin, *Ra'id al-fikr al-masriyy Muhammad Abduh* عثمان امين, رائد الفكر المصري محمد عبده (IslamKotob, 1996).

38 Patricia Owens, *Between War and Politics: International Relations and the Thought of Hannah Arendt* (Oxford: Oxford University Press, 2007), 1–2.

39 Hannah Arendt, *The Life of the Mind* (HMH, 1981), 19–23.

40 Henri Bergson, *Creative Evolution* (Redditch: Read Books Ltd, 2013), 102; Original: "une poussée intérieure qui porterait la vie, par des formes de plus en plus complexes, à des destinées de plus en plus hautes" See: Henri Bergson, *L'évolution Créatrice* (La Gaya Scienza, 2012), 130, http://www.ac-grenoble.fr/PhiloSophie/old2/file/bergson_evolution_creatrice.pdf.

41 George Sorel, *Reflections on Violence* (Cambridge: Cambridge University Press, 1999), 77–79.

42 George Ciccariello-Maher, "To Lose Oneself in the Absolute. Revolutionary Subjectivity in Sorel and Fanon," *Human Architecture: Journal of the Sociology of Self-Knowledge* V (Summer 2007): 101–12.

43 Ato Sekyi-Otu, *Fanon's Dialectic of Experience* (Cambridge, MA: Harvard University Press, 1997).

44 Joan Cocks, *Passion and Paradox: Intellectuals Confront the National Question* (Princeton, NJ: Princeton University Press, 2009), 47.

45 In OoT Arendt uses the term colonial and colonialism more than 80 times.

46 Owens, *Between War and Politics*, 15.

47 Macey, *Frantz Fanon*, 2012, 461–62.

48 Scholarly work would ascertain that Sartre himself would not approve of contemporary terrorism. See Jennifer Ang Mei Sze, *Sartre and the Moral Limits of War and Terrorism* (London; New York: Routledge, 2009), 107–8.

49 Alice Cherki, *Frantz Fanon. Portrait* (Paris: Seuil, 2016), 315.

50 Cherki, 319.

51 Jonathan Fine, *Political Violence in Judaism, Christianity, and Islam: From Holy War to Modern Terror* (Lanham, MD: Rowman & Littlefield Publishers, 2015), 27.

52 For a study in WotE's influence see: Kathryn Batchelor and Sue-Ann Harding, *Translating Frantz Fanon across Continents and Languages:*

Frantz Fanon across Continents and Languages (Abingdon, Oxon: Routledge, 2017).

53 Peter Partner, *God of Battles: Holy Wars of Christianity and Islam* (Princeton, NJ: Princeton University Press, 1997), 239.

54 Cihan Tuğal, "The decline of the legitimate monopoly of violence and the return of non-state warriors," in *The Transformation of Citizenship, Volume 3: Struggle, Resistance and Violence*, eds. Juergen Mackert and Bryan S. Turner (Abingdon, Oxon: Routledge, 2017), 85.

55 Cherki, *Frantz Fanon. Portrait*, 349–50.

56 James Toth, *Sayyid Qutb: The Life and Legacy of a Radical Islamic Intellectual* (Oxford: Oxford University Press, 2013), 304.

57 David Macey, *Frantz Fanon: A Biography* (London; New York: Verso, 2012), 306.

58 Macey, 330.

59 Toth, *Sayyid Qutb*; J. Ward Regan, *Great Books Written in Prison: Essays on Classic Works from Plato to Martin Luther King, Jr* (Jefferson: McFarland, 2015); Juergen Mackert and Bryan S. Turner, *The Transformation of Citizenship, Volume 3: Struggle, Resistance and Violence* (Abingdon, Oxon: Routledge, 2017).

60 Jessica Stern, "The Protean Enemy," *Foreign Affairs*, 1 July 2003, www.foreignaffairs.com/articles/afghanistan/2003-07-01/protean-enemy.

61 Sayyid Qutb, *Milestones* (New Delhi: Islamic Book Services, 2014), 7–13.

62 John Calvert, *Sayyid Qutb and the Origins of Radical Islamism* (Oxford: Oxford University Press, 2009).

63 Tuğal, "The decline of the legitimate monopoly of violence and the return of non-state warriors," 90–91.

64 Fawaz A. Gerges, *Making the Arab World: Nasser, Qutb, and the Clash That Shaped the Middle East* (Princeton, NJ: Princeton University Press, 2018), 200.

65 Jens Hanssen, "Reading Hannah Arendt in the Middle East: Preliminary Observations on Totalitarianism, Revolution and Dissent," *Orient Institut Studies*, no. 1 (2012).

66 For a fascinating account on the Arab world's love and hate toward Sartre see Yoav Di-Capua, *No Exit: Arab Existentialism, Jean-Paul Sartre, and Decolonization* (Chicago, IL: University of Chicago Press, 2018).

67 Idriss Terranti, M. Maddi, and Khaled Boussafsaf, 'La Pensée de Frantz Fanon Dans Le Monde Arabe', Frantz Fanon International, December 24, 2011, accessed March 28, 2019, www.frantzfanoninternational.org/La-pensee-de-Frantz-Fanon-dans-le-monde-Arabe.

68 Alan Hart, *Arafat, a Political Biography* (Bloomington: Indiana University Press, 1989), 123–56.

69 Several authors make that claim directly and indirectly. See: Nick Rodrigo, "Palestine through the Lens of Frantz Fanon," *Middle East Monitor* (blog), October 19, 2015, accessed March 28, 2019, www.middleeastmonitor.com/20151019-palestine-through-the-lens-of-frantz-fanon/; Erica Chenoweth et al., "What Makes Terrorists Tick," *International Security* 33, no. 4 (April 1, 2009): 180–202.

70 Cherki, *Frantz Fanon. Portrait*, 346.

Bibliography

Agamben, Giorgio. *Homo Sacer. Sovereign Power and Bare Life*. Redwood City, CA: Stanford University Press, 1998.

Amin, 'Uthman. Ra'id al-fikr al-masriyy Muhammad Abduh. IslamKotob, 1996. امين, عثمان. رائد الفكر المصري محمد عبده

Arendt, Hannah. *Between Past and Future*. New York: Penguin Classics, 2006.

———. *Crises of the Republic: Lying in Politics, Civil Disobedience, On Violence, Thoughts on Politics and Revolution*. San Diego; New York; London: HMH, 1972.

———. *Eichmann in Jerusalem*. New York: Penguin, 2006.

———. *On Violence*. Orlando; Austin; New York; San Diego; London: HMH, 1970.

———. *The Human Condition*. Chicago, IL; London: The University of Chicago Press, 1998.

———. *The Life of the Mind*. San Diego; New York; London: HMH, 1981.

———. *The Origins of Totalitarianism*. Orlando; Austin; New York; San Diego; London: HMH, 1973.

Batchelor, Kathryn, and Sue-Ann Harding. *Translating Frantz Fanon across Continents and Languages: Frantz Fanon across Continents and Languages*. Abingdon, Oxon: Routledge, 2017.

Benjamin, Walter. *Reflections: Essays, Aphorisms, Autobiographical Writings*. Edited by Peter Demetz. New York: Schocken, 1986.

Bergson, Henri. *Creative Evolution*. Redditch: Read Books Ltd, 2013.

———. *L'évolution créatrice*. La Gaya Scienza, 2012. Accessed June 16, 2019. www.ac-grenoble.fr/PhiloSophie/old2/file/bergson_evolution_creatrice.pdf.

Bernasconi, Robert. "Fanon's 'The Wretched of the Earth' as the Fulfillment of Sartre's 'Critique of Dialectical Reason'." *Sartre Studies International*, no. 2 (2010): 36–46.

Bhabha, Homi K. "Framing Fanon." In *The Wretched of the Earth*, by Frantz Fanon, vii–xli. New York: Grove Press, 2007.

Calvert, John. *Sayyid Qutb and the Origins of Radical Islamism*. Oxford: Oxford University Press, 2009.

Carroll, David. *Albert Camus the Algerian: Colonialism, Terrorism, Justice*. New York: Columbia University Press, 2007.

Caute, David. *Franz Fanon*. Vichy: Seghers, 1970.

Césaire, Aimé. *Discourse on Colonialism*. Translated by Joan Pinkham. New York: Monthly Review Press, 2001.

Chenoweth, Erica, Nicholas Miller, Elizabeth McClellan, Hillel Frisch, Paul Staniland, and Max Abrahms. "What Makes Terrorists Tick." *International Security* 33, no. 4 (1 April 2009): 180–202.

Cherki, Alice. *Frantz Fanon. Portrait*. Paris: Seuil, 2016.

Cheyette, Bryan. "Postcolonialism and the Study of Anti-Semitism." *American Historical Review* 123, no. 4 (October 2018): 1234–45.

Ciccariello-Maher, George. "To Lose Oneself in the Absolute. Revolutionary Subjectivity in Sorel and Fanon." *Human Architecture: Journal of the Sociology of Self-Knowledge* V (Summer 2007): 101–12.

Cocks, Joan. *Passion and Paradox: Intellectuals Confront the National Question*. Princeton, NJ: Princeton University Press, 2009.

Curthoys, Ned. "The Refractory Legacy of Algerian Decolonization: Revisiting Arendt on Violence." In *Hannah Arendt and the Uses of History. Imperialism, Nation, Race, and Genocide*, edited by Richard King and Dan Stone, 109–29. New York; Oxford: Berghahn Books, 2007.

Di-Capua, Yoav. *No Exit: Arab Existentialism, Jean-Paul Sartre, and Decolonization*. Chicago, IL: University of Chicago Press, 2018.

Elon, Amos. "Introduction." In *Eichmann in Jerusalem. A Report on the Banality of Evil*, by Hannah Arendt, i–xxi. New York: Penguin, 2006.

Fanon, Frantz. *Black Skin, White Masks*. Translated by Richard Philcox. New York: Grove Press, 2008.

———. *The Wretched of the Earth*. Translated by Richard Philcox. New York: Grove Press, 2007.

Fine, Jonathan. *Political Violence in Judaism, Christianity, and Islam: From Holy War to Modern Terror*. Lanham, MD: Rowman & Littlefield Publishers, 2015.

Finlay, Christopher J. "Hannah Arendt's Critique of Violence." *Thesis Eleven* 97, no. 1 (May 2009): 26–45.

Gerges, Fawaz A. *Making the Arab World: Nasser, Qutb, and the Clash That Shaped the Middle East*. Princeton, NJ: Princeton University Press, 2018.

Hanssen, Jens. "Reading Hannah Arendt in the Middle East: Preliminary Observations on Totalitarianism, Revolution and Dissent." *Orient Institut Studies*, no. 1 (2012). Accessed March 28, 2019. www.perspectivia.net/publikationen/orient-institut-studies/1-2012/hanssen_hannah-arendt.

Hart, Alan. *Arafat, a Political Biography*. Bloomington: Indiana University Press, 1989.

Hudis, Peter. *Frantz Fanon: Philosopher of the Barricades*. London: Pluto Press, 2015.

Isaac, Jeffrey C. *Arendt, Camus, and Modern Rebellion*. New Haven, CT: Yale University Press, 1992.

Khatib, Sami. "Towards a Politics of 'Pure Means': Walter Benjamin and the Question of Violence." *Anthropological Materialism* (blog), August 28, 2011. Accessed March 28, 2019. https://anthropologicalmaterialism.hypotheses.org/1040.

Lanzmann, Claude. *The Patagonian Hare: A Memoir*. Translated by Frank Wynne. New York: Farrar, Straus and Giroux, 2013.

Macey, David. *Frantz Fanon: A Biography*. London; New York: Verso, 2012.

Mackert, Juergen, and Bryan S. Turner. *The Transformation of Citizenship, Volume 3: Struggle, Resistance and Violence*. Abingdon, Oxon: Routledge, 2017.

Merleau-Ponty, Maurice. *Fenomenologia Percepcji*. Translated by Małgorzata Kowalska and Jacek Migasiński. Warszawa: Fundacja Aletheia, 2001.

———. *Phenomenology of Perception*. London: Psychology Press, 2002.

Migasiński, Jacek. *Merleau-Ponty*. Myśli i Ludzie. Warszawa: Wiedza Powszechna, 1995.

Nayar, Pramod K. *Frantz Fanon*. London; New York: Routledge, 2012.

Owens, Patricia. *Between War and Politics: International Relations and the Thought of Hannah Arendt*. Oxford: Oxford University Press, 2007.

Partner, Peter. *God of Battles: Holy Wars of Christianity and Islam*. Princeton, NJ: Princeton University Press, 1997.

Philcox, Richard. "On Retranslating Fanon, Retrieving a Lost Voice." In *The Wretched of the Earth*, 241–51. New York: Grove Press, 2007.

Qutb, Sayyid. *Milestones*. New Delhi: Islamic Book Services, 2014.

Regan, J. Ward. *Great Books Written in Prison: Essays on Classic Works from Plato to Martin Luther King, Jr.* Jefferson: McFarland, 2015.

Rodrigo, Nick. "Palestine through the Lens of Frantz Fanon." *Middle East Monitor* (blog), October 19, 2015. Accessed March 28, 2019. www.middleeastmonitor.com/20151019-palestine-through-the-lens-of-frantz-fanon/.

Romdenh-Romluc, Komarine. *Routledge Philosophy GuideBook to Merleau-Ponty and Phenomenology of Perception*. London; New York: Routledge, 2010.

Sartre, Jean Paul. *Critique of Dialectical Reason*. Vol. 1. London; New York: Verso, 2004.

Sekyi-Otu, Ato. *Fanon's Dialectic of Experience*. Cambridge, MA: Harvard University Press, 1997.

Sorel, George. *Reflections on Violence*. Cambridge: Cambridge University Press, 1999.

Stern, Jessica. "The Protean Enemy." *Foreign Affairs*, 1 July 2003. www.foreignaffairs.com/articles/afghanistan/2003-07-01/protean-enemy.

Sze, Jennifer Ang Mei. *Sartre and the Moral Limits of War and Terrorism*. London; New York: Routledge, 2009.

Terranti, Idriss, M. Maddi, and Khaled Boussafsaf. "La Pensée de Frantz Fanon Dans Le Monde Arabe." *Frantz Fanon International*, December 24, 2011. Accessed March 28, 2019. www.frantzfanoninternational.org/La-pensee-de-Frantz-Fanon-dans-le-monde-Arabe.

Toth, James. *Sayyid Qutb: The Life and Legacy of a Radical Islamic Intellectual*. Oxford: Oxford University Press, 2013.

Tuğal, Cihan. "The decline of the legitimate monopoly of violence and the return of non-state warriors." In *The Transformation of Citizenship, Volume 3: Struggle, Resistance and Violence*, edited by Juergen Mackert and Bryan S. Turner, 77–92. Abingdon, Oxon: Routledge, 2017.

Zeilig, Leo. *Frantz Fanon: The Militant Philosopher of Third World Revolution*. London; New York: I.B. Tauris, 2016.

2 Violence vs. power

Sartre's preface and the hijacking of Fanon

Fanon's work had become famous also, if not mostly, owing to Sartre's preface. The preface has come to be read as Fanon's rather than Sartre's even though the two are not in agreement or are talking about separate issues. The preface does not leave much room for interpretation – it is an espousal of violence. The text is simply violent – Fanon's book is not, the preface speaks to the European, Fanon's text does not. Cherki, an acquaintance of Fanon's, is of opinion that Sartre "betrayed" Fanon with that preface. To her knowledge Fanon, having read it, conspicuously did not say a word.[1] Sartre's take on violence is different than Fanon's – the humanist thought that Fanon sublimely weaves throughout the pages of WotE suffocates under the heavy body of Sartre's preface. The reception of WotE might have been different if Sartre only reviewed the book in *Esprit* or *Présence Africaine*.

Sartre gives five reasons for the necessity of violence:

1 European hypocrisy and immorality.

The post-Second World War perspective weighs heavily on Sartre's views about Europe, which "never stops talking of man yet massacres him at every one of its street corners, at every corner of the world."[2] "Europe is done for." Low morality of the one against whom violence is used justifies the violence. Sartre saw Europe as ending, declining, dying, because of its immorality. Sartre claims this is also Fanon's diagnosis.

2 European policies that harden divisions and conflicts.

Europe has created the socio-political relations in the colony and between the colony and the metropolis. Violence is the logical response to this policy. "Europe has hardened the divisions and conflicts, forged classes, and in some cases, racism and endeavoured by every means to generate and deepen the stratification of colonized societies." "In order to wage the struggle against us, the former colony must wage a struggle against itself."[3]

3 European wealth.

In the dialectical reasoning if one is well-off, someone else has to be poor, wealth is inseparable from poverty. Europe's wealth has come at the cost of the colonies' poverty. Why would sons and daughters of colonists have to pay for their parents' sins? Because they were the pioneers that made Europe rich. They pushed the limits of indecency so that the majority of the metropolis could remain liberal, humane and culture-loving. To be all these is a privilege of the rich and dominant.

4 European violence.

Europe taught others violence. It is only returning home. It's the same cruelty inflicted on the colonised by the colonist that later the colonist gives back – he understands only the language of violence.[4] It is not *their* violence, it is ours – says Sartre: "in a time of helplessness, murderous rampage is the collective unconscious of the colonized."[5]

5 The reconstruction of man – both the colonist and colonised.

Through violence the oppressed reconstructs himself in that he ceases oppression altogether. This violence, Sartre claims, is man reconstructing himself. Killing a European to him is like killing two birds with one stone: the oppressor and the oppressed. "This is the age of boomerang, the third stage of violence: it flies right back at us, it strikes us and, once again, we have no idea what hit us."[6] Fanon talks *about* the European, but never *to* the European.[7] Yet, thanks to the violence of the colonised Europeans will shrug the burden of colonialism. "We, too, peoples of Europe, we are being decolonized: meaning the colonist inside every one of us is surgically extracted in a bloody operation."[8]

Fanon's violence and power

Fanon's book was not a call to political independence nor was it about violence *for* independence. Independence was happening – many countries formally had already become or were becoming independent. Neither was it about a shortage of violence – violence was enough. But it was about a specific kind of violence: a no choice violence that both ends and creates. WotE analysed the sensitive phase right after independence, the time when the young body of the future nation (not yet a nation) already begins to rot. The disastrous future of the newly independent states on the horizon scared Fanon and made him write WotE. It was the future of a post-colonial world that morphed into today's reality. A similar fear gripped Arendt although the object of her concern was not the colonial world with its specificity but the Western world. Their criticism of the societies they knew will be discussed in the following chapter.

It is striking, given Sartre's preface and the title of the chapter on violence in WotE ("Concerning Violence"), how little in fact Fanon writes about violence per se. Fanon's own definition of violence makes it a fundamental praxis thanks to which social reality poses itself clearly: "violence alone, perpetrated by the people, violence organized and guided by the leadership, provides the key for the masses to decipher social reality" (WotE, 96). The function of violence here is to allow the people, who are devoid of any other means of discerning true social relations, to see reality. The definition of violence is narrowed down by two conditions: (1) it has to be organised and (2) led by a leadership. For the "politically committed," urgent decisions are needed on means and tactics, i.e. direction and organisation. Anything else, Fanon says, is but "blind voluntarism with the terribly reactionary risks this implies" (WotE, 21). In line with Arendt's arguments, if it is not to become blind and uncontrollable, violence has to be **limited**. Fanon's hesitant fear of uncontrolled, unlimited violence is seen in the differentiation between revolutionary and counterrevolutionary brutality. Although revolutionary brutality "hates subtleties and individual cases," the counterrevolutionary brutality is adventurist and anarchist (WotE, 95). If the latter one, which is pure and total, is not contained it will destroy the whole movement (WotE, 95).

What are the social relations in the conditions of colonialism that need to be seen clearly? These are relations between and within two separate compartments: the European and the native. The European and the native sectors in Algeria are irreconcilable and mutually exclusive. There is nothing universal about any one of these worlds. Paved roads; stone and steel; healthy, free and erotic women vs. the medina; the casbah; and piles of trash, dirt and illness. The division between the two compartments resembles today's division between social classes in the Middle East that are impenetrable. In the heart of Cairo two worlds exist side by side: the rich and the poor. At first sight they are both interwoven: the peasant in rugs can be seen in front of a jewellery store but up close these worlds do not mix. The rich drink coffees for five euros in cafes of the Marriott hotel – the coffee of the poor is 20 times cheaper and drank in the dirt of the street. The street should not be seen at all – the rich drive in climatised cars with chauffeurs over highways that run far above the street level. In the street even dogs run away from humans. "The colonist's feet can never be glimpsed" (WotE, 4) – the climatised cars of the rich prevent not only the feet but also the face from being seen. Except those unseen are no longer Europeans; they are Egyptians of the elite. The state galvanises these divisions to facilitate its non-democratic rule.

Colonisation is material expansion on land that causes body failure and dehumanisation. Once the expansion reaches a certain point the colonised defence system collapses. The "generalization of inhuman practices" triggers mental illness. The colonial reality is based on the "systematized negation of the other," on denying the other attributes of humanity, causing the other to question his/her own identity. Not even the Germans dehumanised the occupied French, Fanon would say, although he would not mention the dehumanisation of the Jews (WotE, 182). In the section about mental and psychosomatic disorders Fanon tells several stories of his patients' illnesses: a fighter's wife was raped while he was in combat, but she did not disclose any information and still he is uncertain if he will "take her back"; a fellah survives a massacre and has homicidal inclinations; a man whose mother was killed himself kills a French woman; a French police inspector tortures his wife and children because he had tortured so many natives; and, finally, two Algerian teenagers kill their French friend. "Are you sorry you killed someone?" "No, because they want to kill us" (WotE, 199). The section leads to show how war and national struggle degrade bodies in their mental and physical respects. All is dirty – the colonist, the colonised, the mind, the body, the street. Stomach ulcers, disturbed menstrual cycles, hair whitening, muscular stiffness – abnormality becomes normality.

Why are the colonised unable to see reality clearly and why would it only be violence that lets them see it? Because the colonised themselves are a sick aggregate of beings, without a political appearance – Arendt would append – and without values. Interestingly, the colonist society comes as proportionally sick and ridden with hypocrisy. In this state of relations "it's them or us" (WotE, 42–3), the two Manichean compartments of the world are mutually exclusive. To change it one of the two needs to be removed. The synthesis can only occur when the thesis and the antithesis are no longer existent – a means that destroys the oppressed and the oppressor, a dialectical necessity on the way toward synthesis is searched for here. Since mass suicide of one of the worlds is impossible, it takes demolition of the colonist's sector in the hope that it will also destroy the colonial world. Violence is a door out of an empty room. Violence is the only praxis that both compartmentalised worlds know. Not only does it unite the fighters, it perversely unites everyone on an easy, accessible, understandable platform. It is **universal**.

> To read Fanon is to understand not only the injury that fuels the violence of the native but also the fear that fuels the violence of the settler.[9]

In line with Sartre's argument of violence being merely the **logical** response to the colonist's violence Fanon sees it as "praxis." The violence of the colonised, being the antithesis of the colonial violence, is proportional to it. The Third World counter-violence and the violence of the colonist balance each other out in a proportional "reciprocal homogeneity" (WotE, 43–6). The colonised will give back as much counter-violence as the colonist exerts. It is a cycle: violence causes counter-violence which causes more violence again. Violence is **cyclical**.

The colonist showed the colonised the path to liberation through his own posture. Since the coloniser only understands the language of force (yet hypocritically denies that by claiming to be rational and peaceful), the colonised has to use it, violence is **not optional** – there are no other means of change. It is born the moment the colonised puts a name to all of his misfortunes, and casts all his hatred and rage in this new direction. "They know they are not animals. And at the very moment when they discover their humanity, they begin to sharpen their weapons to secure its victory" (WotE, 7). When this is combined with saber-rattling exercises of the colonist the aggressiveness of the colonised is strengthened. "The gun goes off on its own for nerves are on edge" (WotE, 30–2). As a means it has no alternative although violence is **not self-growing**, it is enhanced by the aggressiveness of the colonist. Arendt makes a distinction between rational and irrational violence – she would consider the one that has one name for all misfortunes as irrational. Fanon's confrontation is not rational. For rationality is of the coloniser. In the phase after the independence the colonist would use rational means, would talk, would discuss – this is not the tool the colonised uses. Because of the irreconcilable Manicheism the tool needs to be different from the colonist's tool who, once violence is no longer useful, would use positivist methods: the rational confrontation, which is violence by other means. For Fanon, the confrontation needs to be irrational. Fanon admits that the violence that may be used (although he does not explicitly speak of violence here, rather of a confrontation) is irrational. Knowingly he stops short of saying that rationality altogether is a Western value or concept.

Who exactly is the enemy in the struggle? Fanon distinguishes between two phases. In the first phase, which lasts until the flames get burning and the independence struggle is in full swing – only the ruling colonist. In the second phase of young independence it is the mythical reality (in the form of religion and feudal tradition), and the assimilated elite.

"The ruling species is first and foremost the outsider from elsewhere, different from indigenous population" (WotE, 5). The criterion of being from outside of the native land overlaps with the wealth criterion: "The cause is effect: you are rich because you are white, you are white because you are rich." It is uncertain if Fanon writes about the colonists meaning all Europeans all just those in the colonised land: "we made this land" (WotE, 14) but the colonist makes and writes history, not anyone else in this world. For Fanon colonialism is a system and a structure, a superstructure that lives beyond the colonists. In Fanon's language it almost becomes a living creature, a behemoth that falls, stands up, pulls strings, takes advantage, snickers – even once the colonists are gone, colonialism is not, its shadow monster lingers on (i.e. WotE, 106). What seems to worry Fanon is that this creature takes hold of the colonised themselves once the colonists are gone. The colonised turns into colonists.

The colonised subject is a persecuted man who is forever dreaming of becoming the persecutor, but who will never be one. Yet, in the second phase of the liberation, which starts when the colonised invariably turns into the colonist, the native can easily be lured into the arms of the metropolis. The colonies have turned into a market and the colonial population are now consumers. The colonised assimilates to the oppressor. What impairs the dialectical development is the colonised intellectual who wants to assimilate to the coloniser and with that serves as a rotten example to others. The colonised intellectual is the element that destroys the logic and slows down the change. On the specific issue of violence, the elite are ambiguous. They are violent in their words and reformist in their attitudes. The colonised intellectual has invested his aggression in his barely veiled wish to be assimilated to the coloniser's world. He has placed his aggression at the service of his own interests. The masses, however, have no intention of looking on as the chances of individual success improve. What they demand is not the status of the colonist, but his place.

The colonist is not the only enemy – perhaps an even greater one is "mythical reality," a reality of magic and religion that allows for the escape from freedom (as in Marx's and Feuerbach's explanations of religion) instead of achieving it. This would also be the primary argument of Fanon's against jihadism – a mythical reality of extraterrestrial justice and solidarity, which in its actuality is nothing more than oppression and hatred. This magical superstructure that permeates the indigenous peoples is characteristic of underdeveloped societies. It shows for example in the fact that libido is primarily a matter for the group and family. Sexual affairs are a community thing, not an individual affair. He explicitly warned "not to perpetuate feudal traditions that give priority to men over women" (WotE, 141). Mythical reality is more dangerous than the colonist – the moment the two merge as one adversary, action against them can be taken. Accordingly and ultimately, when a fissure appeared within the FLN Fanon would side with the secularists against the Islamists.[10]

On the personal level "violence is a cleansing force. It rids the colonised of their inferiority complex" and restores self-confidence (WotE, 51). The sentence is most often quoted to illustrate Fanon's appraisal of violence. But the function of violence as a cleansing force has to do with its praxis feature. Any praxis, as a real action of the body, is a cleansing force. The tragedy of the colonised is that he *only* knows violence and thus has to make it into praxis. Fanon himself seems hesitant about violence and infuriated by the fact that violence is the solution. This hesitancy is seen on the linguistic level. In *Reflections of violence* Sorel uses the word violence almost 400 times, Fanon – 90. The books are comparable in size. In fact he would much rather see "struggle" (used in WotE more than 160 times) as the absolute praxis but struggle has an inbuilt continuity in it and lacks energy and urgency. "Violence," "struggle," "confrontation" are often used as synonyms in WotE.

On the communal level violence unites the people – eliminates regionalism and tribalism (WotE, 50). Individualism in the struggle becomes brotherhood, sisterhood and camaraderie (WotE, 11). Thus violence is a way for a new nation to emerge and demolish the colonial system. Although specifically there are two ways to do just that: violent struggle by newly independent people or outside violence by other colonised people on behalf of all colonised (WotE, 30). Dien Bien Phu was not only a Vietnamese victory – other colonised peoples think and talk about their own Dien Bien Phus. This point is incoherent with the necessity of the struggle to be one's own, nation's own and invites comparisons with jihadist terrorism, which – even if perpetrated by a small group of people – is rhetorically supposed to serve the whole Muslim community. Fanon's expectation that Third World solidarity, including pan-Arab solidarity, will prevail over localities and the struggle by one would mean the struggle by all has, for now, proven false. The most emblematic solidarity cause for the southern European neighbourhood has been the fate of the Occupied Palestinian Territories and their people. Yet the issue has only served most Arab states to reinforce their oppressive ways of governing over their own populations and exculpate their own sins vis-à-vis the people by vilifying Israel and finding an outer enemy.[11] In fact, the tragedy of the most humiliated and degraded Arab peoples today – Palestinians in Gaza, Yemenis and Syrians – is the product, among other reasons, of their brotherly Arab governments. The border crossing in Rafah between the Gaza Strip and Egypt was only open for a dozen days in 2017, adding to the humanitarian catastrophe in Gaza, the Saudis have been bombarding the people of Yemen since March 2015 and the Syrians have been butchered by their own government, their own fellow citizens, other Arabs from Iraq, Libya, Tunisia, Jordan and the Gulf. The state – religious (Saudi Arabia, Iran) or secular (Egypt, Syria) – uses the same structural violence that the colonist had used.

Once independence is there it is illusory because the elite have betrayed the people. "Gabon is an independent country, but nothing has changed between Gabon and France, the status quo continues." The perseverance of the colonialism shows in the economic system and the elite that takes the colonists place, aided by religious superstructure. It becomes a curse for the Third World for it will not stand on its feet not in centuries to come. Money is needed, not moral reparations, Fanon is saying. This money has not been paid: "moral reparation … doesn't feed us" (WotE, 56–7). Fanon describes how party politics are played internally (pluralism dissolves political focus) and externally by the coloniser (reviving tribalism, dividing the people). Colonialism pays the tribal chiefs and *kaids* money for their loyalty – people follow them disciplined. In that period "hatred is not an agenda" (WotE, 89). The colonist will use psychological ploys to diffuse hatred. Experts and sociologists devise these new psychological methods to diffuse it – instead of humiliation they resort to flattery and begin to address the colonised per "sir,' "madam." It becomes quickly forgotten by the colonist that the colonised had been not so long ago accused of being lazy and fatalistic. Experts and expertise is on the side of colonialism, while spontaneity is on the side of the colonised. The vitality of the body is the sole weapon of the colonised if the colonist has both power and knowledge. Without praxis, without violence "all that is left is a slight readaptation, a few reforms at the top, a flag, and down at the bottom a shapeless, writhing mass, still mired in the Dark Ages" (WotE, 96).

Why is Fanon so critical of the second phase of independence struggle? After all it was violence that led to it and, certainly, violence was not lacking. Fanon saw proliferation of violence all around. What was the problem then? That it did not lead to increased power. The concept of power inferred from WotE is not dissimilar to Arendt's. The opinions and the numbers on which power rests, as we will see Arendt claims, are with the masses for Fanon. "In the valleys and in the forests, in the jungle and in the villages, everywhere, one encounters a national authority" (WotE, 83). The people together in the struggle wield power. But Fanon immediately makes statements, to which Arendt objects the most: "the art of politics is quite simply transformed into the art of war. The militant becomes the fighter. To wage war and to engage in politics are one and the same thing" (WotE, 83). Does Fanon really choose war and violence out of the whole spectrum of available political action? Is it a choice in the first place? He seems to be saying that the only political action available to the Arab is war. Yet, rather than it being exciting, it is tragic. Arendt would herself propose that war between nations has no alternative (OV, 6). Fanon's is a war between nations, or between a nation and a quasi-nation. Action in politics, meaning a concerted action by the people, can only be violence, for the masses of the Third World are united by knowing violence only. In the colonial system their power is solely the spontaneous body, which is all the stronger the more people join in. In post-independence phase, when the elite becomes the new coloniser, again it is the concerted bodily struggle that remains for the people as their weapon.

Not that much has changed since 1960s. When Donald Trump declared Jerusalem the capital of Israel late in 2017 the only available political action for the Muslims in response was street protest – not any significant political action or interest on the part of political representatives of the people discontent with that decision. On the occasion of the 70th anniversary of the independence of Israel in May 2018, which to the Palestinians marks 70 years since expulsion of some 700,000 Palestinians from their homes in 1948, Palestinians in Gaza, deprived of basic components of dignity and reduced to material mass of bodies could only use this matter, their own bodies as means of transmitting their political opinion. The result was at least 87 dead and more than 12,000 injured.[12] Arendt would point that "death as an equalizer plays hardly any role in political philosophy" (OV, 68) – in political philosophy it may play no role but for people devoid of death-free, risk-free means of political expression and concerted action, death is as political as it can be.

In the context of generating and organising people's power Fanon discusses the importance of party politics and its leadership. He laughs off its obsession with the question of succession (WotE, 125–27), which vividly reminds of the same debates shortly before the Arab upheavals of 2011 in Egypt, Algeria, Libya, Syria, Yemen.[13] There too the obsession of the ruling parties was who would succeed the leader instead of striving to connect with the people as broadly and directly as possible. The country must possess a genuine party, one that resides in rural areas and is decentralised. Through party people exert their power. The interior should be given priority as a security valve to the disproportional swelling of *pouvoir* in the capital. The masses are capable of discerning the most complex issues. Fanon seems to be saying that a village, a locality has its own common sense that works only in commonality: individuals may not understand an issue, but the totality will (WotE, 130). Jean Jacques Rousseau's concept of "general will" reverberates in Fanon's insights not only on the community but also on force and power.[14]

Intelligentsia, too, needs to be close to the masses. One is either with the masses or is a traitor. This notion effectively disqualifies Fanon as a possible advocate for jihadism – jihadism may make claims to superficial traits of humanism but it disconnects from the masses, it kills the masses in the markets of Baghdad. Political parties are not close to the masses either – in undemocratic corrupt systems they are mostly conglomerates of apparatchiks. But in the young Middle East (more than 60% are less than 25 years of age) a civil society scene is booming despite hurdles thrown by the system in the form of strict NGO-controlling laws. These activist young people follow in the footsteps of Fanon through hard work in NGOs and local cooperatives etc.[15] They resist jihadism precisely with their proximity to the masses.

Fanon's rebuttal of jihadism could also be seen in the insistence of using a language comprehendible to the masses – the language of the Qur'an is anything but clear. Fanon advocates using simple language while talking about complex issues – only then can one appreciate the cognitive powers of the people – if they understand what is being said, they will understand the complexities of the issues discussed. The choice of language reveals the intentions of the speaker (WotE, 130–31). Like Arendt he is preoccupied with language in the service of humanity, cooperative action and respect for the other.

In view of the above it is an oversimplification to claim that Fanon approved of violence although his linguistic style may impose such interpretation. The book is written to the colonised and its intention is to warn about the direction of political changes in the newly independent countries. It is also itself written in a terribly violent context – the French kill Algerians, Algerians kill the French and other Algerians, terrorist attacks proliferate. Despite the fiery language, however, Fanon's arguments are subtle and inexplicit, which becomes particularly ostentatious in comparison with Sartre's introduction. Fanon's violence has two elements: of coercion to it and creation from it. The coercive and terrifying element of it is no less important than the creative one. In a letter of 1955 Fanon would write that the days that were to come were going to be terrible for Algeria: "European civilians and Muslim civilians are really going to take up the gun. And the bloodbath no one wants to see will spread across Algeria."[16] A leading biographer of Fanon's, David Macey, claims that the notion of violence as a cleansing force has not appeared in Fanon's writings until the WotE.[17]

Arendt's violence and power

In OV Arendt mentions Fanon several times but the book is not a direct response to his WotE. It's Arendt's take on violence in relation to Marxism and Hegelianism on the one hand and thinkers drawn by violence in their writings: George Sorel, Vilfredo Pareto, Fanon to some extent only. It may be that her interest in his book resulted from its popularity among black movements in the US. It was the racial element that connected the realities in the US at the end of the 60s with Fanonian French/Algerian/African/Third Worldist thought of a decade earlier.

Arendt objects to Fanon's justification of violence, which she understands as a creative madness that is supposed to push the historical progress forward and create the new man. It has been noted that she misunderstood Fanon,[18] in claiming that he glorified violence, and rather engaged with other thinkers more than with Fanon. However, it cannot be easily inferred from her work that she misunderstood Fanon. She took his work as a pretext to explain her own thoughts about violence and power. What Arendt did write was that the glorification of violence by the students, inspired by Fanon, is only a "hodgepodge of all kinds of Marxist leftovers" (OV, 19). If violence could heal the wound it had inflicted, as Sartre suggested, then revenge would be the cure for all. She did say that the students were inspired by Fanon, although they might have been influenced by Sartre much more and it was them, the students, who misunderstood Fanon, not Arendt.

Like Fanon, Arendt too is interested in relations between nations. These are marred by the cold war, which provides the context for both writers who, in WotE and OV, can be read as theorists of international relations (IR). Arendt assumes that warfare is still the final arbiter in IR because no substitute to it has been found. For Fanon the Third World is the epicentre of the cold war – the prevalence of violence on the global arena justifies the violence of the colonised (WotE, 36) and affiliates them more with the Soviet camp.[19] Arendt also blames the cold war and the nuclear race for the evil that spreads, but – for her – prevalence does not justify the prevalent. Arendt only finds scarce literature on war and warfare. She highlights Clausewitz's war as "continuation of politics by other means" or Engels's violence as accelerator of economic development (OV, 8). In the 60s these claims had proven untrue: the Second World War was followed by another war, the cold one, and not by peace. Yet the future is unpredictable – all futuristic predictions are only true if nothing of importance ever happens (OV, 7). Human action is so creative and disruptive that predictions must fail. Theories put to sleep our common sense because they indeed are coherent as opposed to a world created by human action, which a priori cannot be neither known nor coherent. Fanon would agree. His violence as praxis or – better – struggle is action in Arendtian sense.

For Arendt violence cannot be an end it itself because it needs implements (or tools), which differentiates it from power, strength and force. Violent action is specific in that the means to achieve the goal overwhelm the end result. The end of human action cannot be reliably predicted – therefore the means to achieve political goals are often of greater relevance than the intended goals/intentions. "Violence harbours within itself an additional element of arbitrariness" (OV, 4). Good or ill luck decide the results of violence, it's uncontrollable even if complex game theory algorithms are applied.

In an introduction to her concept of power Arendt brings up the mistaken classical understanding of it used by communist leaders. They saw power that "grows out of the barrel of a gun" (Mao Tse-tung), which is un-Marxian in her view. Marx taught that violence was a necessary phase before a new society could be born but violence was not means in itself, it was the result of inherent contradictions in the old class society, and it was transient. Violence did not bring about the new society – the disturbed class relations were bound to bring it. Sorel, Sartre and Fanon thought of it differently she claims (OV, 12). Sorel tried to combine Marxism and Bergson's philosophy of life but ended up marrying existentialism and Marxism, just like Sartre. She observes that Sartre glorifies violence much more than Fanon, more still than Sorel in *Reflections on violence*. Arendt makes sure she explains how far from Marx Sartre has gone in his writings on violence, in between the lines implying that Fanon diverged from Marx too. Although in at least one aspect Fanon's idea is still Marxist: the inhumanity of the old relations between the metropolis and the colony resembles the inhumanity of class relations – only after the struggle will a new society, a new humanity emerge.

For Arendt violence also is cyclical although the motor of cycles is technological development rather than dialectics. The New Left grew under the shadow of the atom bomb, which pushed it to create the nonviolence Left. But in 1960s the nonviolence movement was on the defensive. The student rebellion of March 1968 embraced violence again. Since 1968 Arendt sees it as a global phenomenon (OV, 14). No single underlying factor united the causes except for one: "technological progress is leading in so many instances straight into disaster." Here Arendt shares Fanon's criticism of Western culture that is walking towards the edge. The young live with a back thought of a world that soon may not exist. In the context of violence and power she is interested in the way black students, the Black Power movement, use violence. In European campuses violence remained a matter of theory but black students used it in practice, and they could count, unlike European students, on popular support outside of campuses. Controversially, she calls them an "interest group" that was "admitted without academic qualification" and whose interest is to "lower academic standards" (OV, 18). In derogatory terms she mentions the takeover of the Willard Straight Hall at Cornell University in April 1969, which today, in the Cornell chronicle, is described as a takeover that "symbolized an era of change:" "Within Cornell, the takeover has come to be seen as an event that gave birth to enormous social, governance and ideological change. In fact, institutional change was already under way."[20] Her opposition to political decisions that were cornerstones in the progress of racial issues in the US cast a shadow over her advocacy for equality in the political domain while accepting discrimination in the social one. The criticism of Arendt went as far as calling her a white ignorant.[21] Judith Butler thought Arendt's criticism of Fanon "intemperate" and blamed it on her Eurocentrism outright.[22] Indeed, even with a good deal of reverence and distance to socio-political reality one cannot leave her *Reflections on Little Rock* or statements such as "a police force that gives me the creeps, speaks only Hebrew and looks Arabic. (…) And outside the doors, the oriental mob, as if one were in Istanbul or some other half-Asiatic country"[23] unnoticed. As Butler notes,[24] for Arendt dispossession does not apply to racial minorities – in fact, race does not exist as a philosophical or political problem. It makes a fair appraisal of Arendt's take on the violence at US campuses as compared to the 1968 revolt in Europe difficult if not impossible and takes away from the credibility of Arendt's statements about them. Kathryn T. Gines's writings,[25] most notably her book *Hannah Arendt and the Negro Question*, expose how underdeveloped and inconsistent with thoughts on otherness etc. Arendt's stance on racial issues was.

> I am shocked (...) by her casual relegation of racial discrimination (...) outraged by her condescending and stereotypical characterizations of people of African descent.[26]

It may be that her undeveloped stance on racial issues disproportionately widens the disagreement with Fanon on violence, power and agency where there would be little disagreement otherwise, which will be discussed in the following chapter.

Arendt refers to Fanon's WotE as "irresponsible grandiose statements" (OV, 20) but she finds him "closer to reality than most." In a footnote she reveals her sympathetic understanding of Fanon's:

> I am using this work because of its great influence on the present student generation. Fanon himself, however, is much more doubtful about violence than his admirers. It seems that only the book's first chapter, 'Concerning Violence,' has been widely read. Fanon knows of the 'unmixed and total brutality [which], if not immediately combated, invariably leads to the defeat of the movement within a few weeks.'
>
> (OV, 107)

But in the main body of the text she seems to read him as an orthodox Marxist and faults him with misunderstanding Marx. Fanon's wretched dream but, Arendt remarks, for Marx dreams never come true. Marx's implosion of capitalism was a matter of economic laws, the coming of socialism was not a dream but a necessity. Arendt seems to be implying that because of Fanon's inability to handle the terrible reality around him, he escaped into the dream. Slave rebellions are rare and when they happen their "mad fury" turns dreams to nightmares for all. Violent outbursts can only change personnel but not reality. This passage in OV becomes particularly interesting if we think of Fanon's slave rebellion as Sorel or Benjamin thought of a general strike. The slave rebellion, like the general strike, is not a means to an end because there are no concessions to be made and therefore it is a pure means and nonviolent[27] – the slave will not be happy with a half-slave status. The violence of the native destroys all the theoretical consequences of every possible colonist policy just like "the general strike destroys all the theoretical consequences of every possible social policy."[28] Fanon's "mad fury" is Benjamin's divine violence that interrupts application of law to bare life, in the words of Khatib.[29] Even though the thesis deserves a closer and separate look it already shows how distant Arendt becomes in her understanding of violence from Benjamin, her close philosophical companion, and how close to him Fanon may be read.

The agreement on the Left and Right seems to have it that violence is the most flagrant manifestation of power (OV, 35). Which is strange for Arendt because this opinion consents to the state being a coercive superstructure with Weber's monopoly on force, meaning asserting one's own will against the resistance of others (OV, 36). If it were true, Arendt concludes, there would be no difference between a democracy and an autocracy – power in both would be the power of a gun.

To the faulty tradition of understanding violence Arendt juxtaposes the Greek concept of power and law that is not based on the command-obedience relationship but on equality (OV, 40; Table 2.1). The 18th century revolutions borrowed from this tradition to build a rule of law stemming from the power of the people. "It is people's support that lends power to the institutions of a country" – "All governments rest on opinion" she quotes Madison. Institutions are materialisations of power. They decay the moment people cease to support them. Power needs numbers (statistics), while violence needs implements, or agents, or instruments (OV, 41). Among the implications of her understanding of power is that majority rule can be an effective system to suppress the rights of minorities and dissent without using violence. That is why "the extreme form of power is All against One, the extreme form of Violence is One against All."

Power, strength, authority, force, violence are not synonyms. Behind the mistake of understanding them synonymously is one preoccupation: with who rules whom. These words could be synonymous if public affairs were a business of domination, which they are not. Arendt gives definitions to these words. Power is the ability to act in concert, strength is a property of the individual, force is reserved for the force of nature or circumstances, authority is personal: a father has authority over a child, a church over a believer (through giving absolution) and does not entail coercion (father can lose authority by beating the child). Violence is distinguished by its instrumental character, as such it is close to strength.

It is common to see violence and power together (OV, 46). If we think of power in terms of command and obedience, then it can equate

Table 2.1 Features of the traditional and Arendtian understandings of power

	Manifestation	*Basis*	*Prerequisite*	*Features*	*Result*
Common understanding of power	Violence	Command	Obedience	Distant/ divided	Temporariness
Arendt's understanding of power	Action in concert	Concession	Support	Proximate/ approaching/ united	Immortality

with violence. When governments are challenged, they need violence to keep power and it then becomes clear that power was only a disguise for violence. But it is not plausible to think this. Even totalitarian rulers need a power basis, none can be solely based on violence, nor can a single man use violence successfully.

According to Arendt, Alexander Passerin d'Entreves, an Italian philosopher and historian of Italian law (d. 1985), is the only author who does see the distinction between power and violence. In *The Notion of the State* he observes that power is qualified violence or institutionalised violence. But Arendt goes further. For her the distinction is total because violence is the opposite of power (OV, 55).

Violence is by nature instrumental and needs justification. All that needs justification cannot be the essence of anything. Violence is not the essence of power, strength or force. On the contrary, concepts such as power and peace are ends in themselves. Power does not need justification (it is inherent in political affairs) but it does need legitimacy. Legitimacy happens when people act together, it precedes power, it is in the past. While justification refers to a goal in the future.

"Violence can be justifiable but it never will be legitimate" (OV, 52). The further in the future the justification finds its source, the less plausible it is. Immediate justification, as in self-defence, is plausible. A justification of violence that comes late or never is doubtful or cannot be qualified as justification at all. The war in Iraq of 2003, which was justified by the US on the basis of Iraq possessing weapons of mass destruction, became unjustifiable when those weapons had not been found. Alternative justifications such as the need to democratise an authoritarian system surfaced almost a year into the occupation of Iraq and failed to convince the international public opinion.

The textbook case of a confrontation between violence and power that Arendt gives is the Russian army and the nonviolent resistance of the Czechoslovaks in 1968 (OV, 52). The less power the more violence and the more temptation to substitute violence for power. Russia had little power, so it struck Czechoslovakia with greater violence. If it wasn't England that responded to the powerful Ghandi but Russia the result would not be decolonisation but submission and massacre. Violence can destroy power but it will not turn into power.

For Arendt power is the exact opposite of violence (Figure 2.1), with diametrically different qualities (Table 2.2). Where one rules

Figure 2.1 Total opposition of violence to power.

Table 2.2 Arendtian characteristics of power as opposed to violence

	Sine qua non	*Requirement*	*Status*	*Relation to politics*	*Relation to one another*	*System of government*
Power	Legitimacy –> precedes power	Support/ numbers/ opinions	End in itself	Essentially political	Doesn't need violence	Participatory democracy
Violence	Justification –> succeeds violence	Implements	Means to achieve the goal, overwhelm the result	Uncontrollable (chance decides)/ anti-political	Destroys power	Terror

absolutely, the other is absent (OV, 55). The opposite of violence is not nonviolence. In the case of the violence-power opposition dialectics (Hegel, Marx) that believe in the advancement through the power of negation[30] are useless, for violence cannot create power, it can only destroy it. She implies that in this way Fanon uses dialectics to inspire "treacherous hope" and dispel "legitimate fear" (OV, 58). According to Arendt, Fanonian dialectic is as treacherous as the Hegelian and Marxist dialectic has proven to be in relation to violence. Violence is not merely a necessary phase, a tool, a means – it is the anti-thesis of power. It is not evil, however, for human violence is human specific.

Rational and irrational violence

Arendt differentiates between rational and irrational violence. She points to rage as the reason for violence. Her rage is what Fanon would call anger in WotE. Rage can be irrational. In dehumanised conditions (i.e. concentration camps) people do not necessarily become violent, animal-like and rage, but vice versa – their dehumanisation can be seen in the absence of rage. Rage is not an automatic reaction to misery (earthquake, disease), it transpires when our sense of justice is offended, when something can be done about misery but is not. Violence can set the scales again in certain circumstances, specifically it can immediately repair obvious injustice. Arendt gives the example of Billy Budd, the title protagonist of Herman Melville's novel, who kills the man that falsely testified against Budd[31] (OV, 64). Arendt finds his behaviour unconstitutional and anti-political but human. In WotE Fanon gives an analogical literary example of the protagonist of Aimé Césaire's play *And the dogs were silent* who kills his oppressor, the white master. "I struck, the blood spurted: it is the only baptism that today I remember" (WotE, 46). According to Arendt Billy's violence is rational and so has to be Césaire's, although Arendt does not refer to it.

Emotions and rationality are neither inversely conditional nor are they antonymous. Violence is irrational when it takes substitutes for subjects. The rebel in Aimé Césaire's play rages not against a substitute but against his real oppressor, so does Billy Budd. "Rage and violence turn irrational only when they are directed against substitutes" (OV, 64). Substitutes are those who are not the real culprits. The example of irrational violence that Arendt gives is the black rage against whites, which was stirred by whites themselves with their "We are all guilty" approach – for Arendt it is racism in reverse. If, absurdly, all black are innocent and all white are guilty no reconciliation is possible. The black rage leads to irrationality. Yet it begs the question Arendt does not pose or answer about when and how does a real culprit turn a substitute culprit? Ultimately, perceptions decide about the real culprits, not bookkeeping. If politics is the immortal sphere of concerted action, then its temporal continuity makes all involved – in the past and present – responsible actors at all times.

Historically it was not so much injustice but rather hypocrisy that caused rage. Fanon, Sorel, Pareto glorify violence for its sake because they are enraged by hypocrisy of the bourgeois society. They are more radical than "conventional Left," inspired by compassion and desire for justice. Why would hypocrisy cause rage and violence? Because it is a trap for reason: "Words can be relied on only if one is sure that their function is to reveal and not to conceal" (OV, 66). Hypocrisy causes alienation from political institutions and is a particular feature of European liberal bourgeois society. Hypocrisy must have made a mockery of human rights and other values in the eyes of everyone: "The very phrase *human rights* became for all concerned – victims, persecutors, and onlookers alike – the evidence of hopeless idealism or fumbling feeble-minded hypocrisy" (OoT, 269). It was this hypocrisy that roused the elite after the First World War to notice the ruin of "fake security, fake culture and fake life" through "purification" in war as Thomas Mann had it and as Arendt quotes him (OoT, 327).

Even if violent reaction to hypocrisy is rational it loses its rationality when it becomes a strategy in its own right – that is when it turns into a hunt for suspects and ulterior motives. It becomes a substitute for a rational strategy to get rid of hypocrisy and to name things correctly. Violence is rational as long as it's effective in reaching the end that justifies it (OV, 78). We do not know the results of an action only until after it, especially in the short term. The risk of violence is that the means overwhelm the end. Action is irreversible and status quo ante unachievable. Violence cannot, therefore, promote causes such as revolution, progress etc. It only "dramatize(s) grievances and bring(s) them to public attention." Arendt does admit that more often violence can be a "weapon of reform than of revolution" (OV, 79). Student riots lead to reform. The goals need to be achieved rapidly or else the result will be the contamination of politics with violence altogether. Violence simply changes the world into a more violent one (OV, 80).

Employing rationality in the analysis of political violence in Fanon's and Arendt's thought leads to inconsistencies. It can be assumed that by introducing the differentiation into rational and irrational violence Arendt makes a point that Fanon's violence should be understood as the latter one because it is directed at substitutes: all colonists, all whites, all pied noirs are enemies. Indeed, such understanding of violence can be inferred from Sartre's preface, but not from WotE's proper text, which differentiates two stages of national liberation with different enemies and different functions of violence. Additionally, if Arendt's example of Billy Budd's killing is rational violence then it should also be valid for Fanon's literary example of the protagonist in Césaire's poem and any kind of violence that takes the structural violence of the colonist state for enemy.

Notes

1 Alice Cherki, *Frantz Fanon. Portrait* (Paris: Seuil, 2016) 319–20.
2 Jean Paul Sartre, "Preface," in *WotE* (New York: Grove Press, 2007), XLIV.
3 Sartre, XVI.
4 Sartre, L.
5 Sartre, LII.
6 Sartre, LIV.
7 Sartre, XLV.
8 Sartre, LVII.
9 Mahmood Mamdani, *Good Muslim, Bad Muslim: America, the Cold War, and the Roots of Terror*, (New York: Harmony, 2005), 10.
10 Francis Jeanson, *Sartre and the Problem of Morality*, Studies in Phenomenology and Existential Philosophy (Bloomington: Indiana University Press, 1980), XX.
11 Such a policy process I have analysed in my doctoral dissertation *Pseudo-stabilization. Problems of contemporary American policy in the Middle East,* published in Polish. See Patrycja Sasnal, *Pseudostabilizacja. Problemy wspolczesnej polityki USA na Bliskim Wschodzie* (Warszawa: Wydawnictwo Naukowe PWN, 2016).
12 "Force Used against Protestors in Gaza 'Wholly Disproportionate' Says UN Human Rights Chief," *UN News*, May 18, 2018, accessed March 28, 2019, https://news.un.org/en/story/2018/05/1010142.

13 Roger Owen, *The Rise and Fall of Arab Presidents for Life: With a New Afterword* (New Haven, CT: Harvard University Press, 2014).
14 See Jane Anna Gordon, "Of Force, Power, and Will: Rousseau and Fanon on Democratic Legitimacy," in *Living Fanon: Global Perspectives*, ed. Nigel C. Gibson, Contemporary Black History (New York: Palgrave Macmillan US, 2011), 201–11; Jane Anna Gordon, *Creolizing Political Theory: Reading Rousseau through Fanon* (New York: Fordham University, 2014).
15 For a preliminary as well as case-focused analysis of the new power of the civil society in the Arab world see Benoit Challand, "The Counter-Power of Civil Society and the Emergence of a New Political Imaginary in the Arab World," *Constellations* 18, no. 3 (September 26, 2011): 271–83; Mohamed 'Arafa, "The Tale of Post-Arab Spring in Egypt: The Struggle of Civil Society Against a Janus-Faced State," *Indiana International & Comparative Law Review* 27, no. 1 (January 2017): 43–78; E. K. Bribena, "Civil Society and Nation Building in the Arab Spring: A Focus on Libya," *Gender & Behaviour* 15, no. 2 (May 2017): 8994–9004.
16 David Macey, *Frantz Fanon* (London; New York: Verso, 2012), 270.
17 Macey, 292.
18 Macey, 470.
19 Unlike Sartre Fanon does not praise the Soviet camp and is not explicit about the necessity of the Third World to ally itself with it. In his opinion, however, the Third World is not excluded from the atmosphere of violence. That is why the colonized understands Khrushchev when he hammers his shoe on the table or Castro when he enters the UN in a military uniform.
20 George Lowery, "A Campus Takeover That Symbolized an Era of Change | Cornell Chronicle," *Cornell Chronicle*, April 16, 2009, accessed March 28, 2019, http://news.cornell.edu/stories/2009/04/campus-takeover-symbolized-era-change.
21 Michael D. Burroughs, "Hannah Arendt, 'Reflections on Little Rock,' and White Ignorance," *Critical Philosophy of Race* 3, no. 1 (2015): 52–78.
22 Judith Butler, "I Merely Belong to Them," *London Review of Books*, May 10, 2007.
23 Hannah Arendt and Karl Jaspers, *Hannah Arendt/Karl Jaspers Correspondence, 1926–1969* (San Diego, New York, London: Harcourt Brace Jovanovich, 1992), 435.
24 Butler, "I Merely Belong to Them."
25 Kathryn T. Gines, "Race Thinking and Racism in Hannah Arendt's 'The Origins of Totalitarianism,'" in *Hannah Arendt and the Uses of History. Imperialism, Nation, Race, and Genocide*, ed. Richard King and Dan Stone (New York, Oxford: Berghahn Books, 2007), 38–53; Kathryn T. Gines, *Hannah Arendt and the Negro Question* (Bloomington: Indiana University Press, 2014).
26 Gines, *Hannah Arendt and the Negro Question*, 2014, xii.
27 Sami Khatib, "Towards a Politics of 'Pure Means': Walter Benjamin and the Question of Violence," *Anthropological Materialism* (blog), August 28, 2011, accessed March 28, 2019, https://anthropologicalmaterialism.hypotheses.org/1040.
28 Sorel, *Reflections on Violence*, 126.
29 Khatib, "Towards a Politics of 'Pure Means': Walter Benjamin and the Question of Violence."

30 According to the dialectical "prejudice," Arendt notes, evil is a private modus of good, it carries the promise of future good.
31 In the book Billy strikes dead John Claggart, a master-at-arms, an officer responsible for law enforcement, who falsely testified against him.

Bibliography

'Arafa, Mohamed. "The Tale of Post-Arab Spring in Egypt: The Struggle of Civil Society against a Janus-Faced State." *Indiana International & Comparative Law Review* 27, no. 1 (January 2017): 43–78.

Arendt, Hannah. *On Violence*. Orlando; Austin; New York; San Diego; London: HMH, 1970.

———. *The Origins of Totalitarianism*. Orlando; Austin; New York; San Diego; London: HMH, 1973.

Arendt, Hannah, and Karl Jaspers. *Hannah Arendt/Karl Jaspers Correspondence, 1926–1969*. San Diego; New York; London: Harcourt Brace Jovanovich, 1992.

Bribena, E. K. "Civil Society and Nation Building in the Arab Spring: A Focus on Libya." *Gender & Behaviour* 15, no. 2 (May 2017): 8994–9004.

Burroughs, Michael D. "Hannah Arendt, 'Reflections on Little Rock,' and White Ignorance." *Critical Philosophy of Race* 3, no. 1 (2015): 52–78.

Butler, Judith. ""I Merely Belong to Them"". *London Review of Books*, May 10, 2007.

Challand, Benoit. "The Counter-Power of Civil Society and the Emergence of a New Political Imaginary in the Arab World." *Constellations* 18, no. 3 (26 September 2011): 271–83.

Cherki, Alice. *Frantz Fanon. Portrait*. Paris: Seuil, 2016.

Fanon, Frantz. *The Wretched of the Earth*. New York: Grove Press, 2007.

"Force Used against Protestors in Gaza 'Wholly Disproportionate' Says UN Human Rights Chief." *UN News*, May 18, 2018. Accessed March 28, 2019. https://news.un.org/en/story/2018/05/1010142.

Gines, Kathryn T. *Hannah Arendt and the Negro Question*. Bloomington: Indiana University Press, 2014.

———. "Race Thinking and Racism in Hannah Arendt's 'The Origins of Totalitarianism'." In *Hannah Arendt and the Uses of History. Imperialism, Nation, Race, and Genocide*, edited by Richard King and Dan Stone, 38–53. New York; Oxford: Berghahn Books, 2007.

Gordon, Jane Anna. *Creolizing Political Theory: Reading Rousseau through Fanon*. New York: Fordham University, 2014.

———. "Of Force, Power, and Will: Rousseau and Fanon on Democratic Legitimacy." In *Living Fanon: Global Perspectives*, edited by Nigel C. Gibson, 201–11. Contemporary Black History. New York: Palgrave Macmillan US, 2011.

Jeanson, Francis. *Sartre and the Problem of Morality*. Studies in Phenomenology and Existential Philosophy. Bloomington: Indiana University Press, 1980.

Khatib, Sami. "Towards a Politics of 'Pure Means': Walter Benjamin and the Question of Violence." *Anthropological Materialism* (blog), August 28, 2011. Accessed March 28, 2019. https://anthropologicalmaterialism.hypotheses.org/1040.

Lowery, George. "A Campus Takeover That Symbolized an Era of Change | Cornell Chronicle." *Cornell Chronicle*, April 16, 2009. Accessed March 28, 2019. http://news.cornell.edu/stories/2009/04/campus-takeover-symbolized-era-change.

Macey, David. *Frantz Fanon: A Biography*. London; New York: Verso, 2012.

Mamdani, Mahmood. *Good Muslim, Bad Muslim: America, the Cold War, and the Roots of Terror*. New York: Harmony, 2005.

Owen, Roger. *The Rise and Fall of Arab Presidents for Life: With a New Afterword*. New Haven, CT: Harvard University Press, 2014.

Sartre, Jean Paul. "Preface." In *Wretched of the Earth*. New York: Grove Press, 2007.

Sasnal, Patrycja. *Pseudostabilizacja. Problemy współczesnej polityki USA na Bliskim Wschodzie*. Warszawa: Wydawnictwo Naukowe PWN, 2016.

Sorel, George. *Reflections on Violence*. Cambridge: Cambridge University Press, 1999.

3 New humanism

> For Europe, for ourselves, and for humanity (...) we must (...) endeavor to create a new man.
>
> Frantz Fanon (WotE, 240)

Fanon and Arendt, like perhaps most philosophers since, think and write in the same historical context: of what highly civilised Europe was capable of doing to itself and to others. The climax of the inward European evil came with the Second World War and the essence of the outward European evil – with colonialism.

Fanon and Arendt read and witness history specifically. Faced with the crimes in Algeria Fanon seems to disregard or at best look past the scale of the human fall in the Second World War, while Arendt, having lived and examined the atrocity of that war, perhaps does not know enough about the tragedies of the Third World. Until today for many intellectuals in the West the history of the south, the Arab world most specifically, is little known, and its violent part even less. Yet both – the Second World War and colonialism, or rather colonialism and the Second World War to be chronologically and, in part, causatively correct[1] – are two faces of the same sophisticated and advanced Europe which produced and used terror against itself and others alike. The "self-destructive" part of the European folly dominates Arendt's world of thought and the "others destroying" part of it guides Fanon's.

The **Second World War** (a well-defined and better-known human period in time) and **colonialism** (a longer, multifaceted and still obscure to the wider public process) do not compare well unless one can find a narrowed down, illustrative quintessence of colonialism. Historians agree that French Algeria is just that.[2] No other colonial territory was so traumatically tied to the metropolis and, perhaps, no other shows the linkage between the Second World War and colonialism so vividly albeit symbolically. It was on 8 May 1945, the Victory in Europe day that demonstrations calling for Algerian independence broke out in Sétif, Eastern Algeria. They were brutally quelled. In the days that followed more than a hundred French settlers were killed. French military responded disproportionately punishing collectively whole villages and districts, killing in and around Sétif and Guelma around 20,000 people, according to French historians,[3] or more than twice this number according to the Algerian sources. The pattern of violence was to be repeated. In Constantine and Philippeville in late August 1955 the FLN killed dozens of French and loyalist civilians, which was followed by a massacre of more than 7,000 Algerian civilians at the hands of the French army in less than a week. On 20 August 1955 thousands of men were herded into a football stadium in Philippeville where the French machine-gunned hundreds of them. Colonial violence also erupted in the metropolis. In October 1961, a couple of months before Fanon's death, in street riots and their aftermath more than a hundred Algerians were murdered by police in Paris and some 14,000 detained. Maurice Papon, the prefect of the Parisian Police was responsible for the death of those people.[4] It comes as little surprise that in 1956–58 Papon was the prefect of Constantine in Algeria. Himself and his conservative milieu symbolised the attitude to people of colour and strangers: violent racism. Papon is not only notorious for the events of 1961: in the Second World War as the deputy to the prefect of Bordeaux under Vichy, he personally sent 1,500 Jews to concentration camps.

In fact, the correct chronological trajectory starts with Maurice Papon as a committed bureaucrat under Vichy who, like Eichmann, sends people to death and masters methods of invigilation, prosecution, brutal repression of the Jews. Under Vichy in 1940 Arendt, too, is sent as a public enemy to the camp in Gurs, less than 200 km south of Bordeaux where Papon is stationed. What he used on Jews in France he would later use on Arabs in Algeria and Morocco. People like Papon – loyal anti-communists – survived the war because they were an important asset in the cold war.[5] Up to a point in time the practices of violence that he implemented in the colonies could not be outright used in France. Not until xenophobic ideology spread enough to let them into Paris:

> These practices, they [historians] argue, were justified by a fully developed ideological program that combined a model of civilizational conflict with elements of social Darwinism, orientalist conceptions of the "Islamic personality," and a crude behavioralist vision of human nature that led to organized "psychological" warfare against entire populations.[6]

Papon epitomises all that torments both Arendt and Fanon: a thoughtless, unreflective European agent of violence that can only spread violence. Responsible for both, he embodies the odious experience of the Second World War and colonialism: two (geographical) sides of the same fascism. From this one and the same root of the European fall sprouts Arendtian and Fanonian search for the new humanism. Repulsed by the enormity of violence they both dream of a new man, an anti-Papon. Arendt finds him in the north, in the public space taken from the ancient Greeks. Fanon is unable to find him in the overwhelmingly racist north, so he puts his hope in the south.

Arendt and Fanon differ in opinion about events in Algeria. France – Arendt writes in between the lines – showed restraint in Algeria. Having lived what he had, Fanon would find this opinion grotesque. He saw the colonial misery and drama first hand. It cannot be known how closely Arendt followed events in North Africa, although in the war in 1943 she wrote a well-informed article about Jews in Algeria and the abrogation of the Crémieux decree, in which she found that French anti-Semitism and the power of the dictatorial French colonists drove politics in Algeria.

> Anti-native rule by the French colonials is possible because of the inferior political status of the natives. (…) French colonials (…) admired Hitler's racial policy and (…) were only too glad to see the violent feelings of the economically depressed and politically underprivileged population directed against Jews rather than themselves.[7]

Based on the findings of Isaac and Curthoys she may have perceived the war in Algeria in 1950s through Albert Camus's lens (as opposed to Sartre), with whom she might have even shared the outrage at the conditions of living of the colonised but under no conditions could she (and neither could he) accept their violence.[8] One might also assume that after the Holocaust nothing could have been as morally debilitating for her, which would explain her disinterest in the world outside of the West.[9] For Fanon, who thought of Jews as brothers in misery and humiliation, the Second World War and the Holocaust were not exceptional: the colonised would live the same, if not worse, fate. The killing, stretched over time, was slower but more persistent, widespread, sophisticated and quiet. Until today and unlike the Holocaust the history of European southern neighbours and their misery is not taught at European schools – not only in France or the UK but all the less so in the "newer Europe" like Poland. The wider public still knows very little about the standard of living of a Muslim Algerian after the Second World War: "four walls of dried mud, a sheepskin on an earthen floor, the bread and the figs that provide the calories needed for physical survival and a ten-hour working day. Shanty towns, people living in caves, malnutrition, filth, illiteracy."[10] Algeria, where Europeans would never shake the hand of an Arab, never drunk a coffee with an Arab was "one immense wound."[11] Maurice Merleau-Ponty calls French politics in Algeria fascism out right.[12] The drama of the Third World is a little known historical fact and Fanon is factual about it. When writing about Europe, however, Fanon does not employ the same factual reading. He seems to loath Europe for its exploitation of the Third World yet admires its achievements if it wasn't for the hypocrisy. In contrast to the Third World European states progressed simultaneously, without the superior-inferior relation, he says. "No nation really *insulted* the others." Polemically although justifiably he treats Europe – like Europeans tend to do of the Middle East or the Arab world – as a homogenous unity. Actually, in WotE he is preoccupied with Europe more than with the Third World. That preoccupation can be seen in his anti-European tirades – "European opulence is a scandal," "Europe is literally the creation of the Third World" (WotE, 53, 58) – which becomes incoherent when he suggests Europe has done a good job and "let us stop accusing it" (WotE, 237). His conclusion is a call to abandon the European model, which only talks of man but, in reality, butchers men. "Let us endeavour to invent a man in full, something Europe has been incapable of achieving" (WotE, 236). To start over a new history of men is the destiny of the Third World: "walk in the company of man, every man, night and day, for all times," "We must be pioneers and create a new man" (WotE, 238, 240). The Third World cannot and must not become Europe but it can and must achieve what Europe wanted but failed to achieve.

Arendt, too, calls for a new man by giving politics its initial meaning and leaning on the concerted action of the people, and, like Fanon, she is haunted by violence in the background. Violence has become a global phenomenon, she writes, although warfare has lost its "glamour." Now deterrence is the guarantor of peace. She is petrified by the technological advancement of nuclear weapons – in the 60s large-scale thermonuclear bombs were tested by the US and USSR.

The similarities between Fanon and Arendt can be seen not only in the impact of the horror of violence that pushes them to work but also, counterintuitively, in the diagnosis and the treatment of the problem. Both are philosophers of responsibility and commitment and both see human act (Fanon) or action (Arendt) as a necessary vehicle of a man creating himself. Both see a nation as formed by bonds of joint action rather than built on xenophobic community of blood. Finally, their goal is the same, as Frazer and Hutchings note, which is to explore the possibility of "politics without violence."[13]

Responsibility and commitment in the political sphere

Arendt and Fanon are intellectuals who feel the pressure of responsibility and who commit to alerting about the dangers they see in the social and political processes.

Fanon is committed to the "decisive struggle to steer the nation in a healthy direction" (WotE, 121). The societal renewal, or even the creation of a new society, is his responsibility and commitment. Why is it important for Third World societies that had already become independent? Fanon is cautioning that precisely because of the independence the dangers may be obscured more than they were prior to the independence struggle. Out of the two stages – the period of the struggle for independence and the period of the newly independent self-governance – the second one is more dangerous and demanding. In the first phase the masses are united with the elite, there is one political faction – a faction for independence – unity of goals is near absolute. In the second phase that unity disintegrates, largely owing to the betrayal of the masses by the elite or, in Fanonian terms, the bourgeoisie. For him the bourgeoisie is not only a social class of owners of the means of production but also of the intelligentsia. In post-independence period the new rulers, politicians, businessmen – even though they hail from the local ethnicity – they simply take the place once held by the colony. The ruling elite gets a new face but retains the ways of the colonist. The function of violence in the two periods then has to be different: it is a necessity in the first phase but it becomes a dangerous "shortcut" in the second. If the text of WotE becomes at times incoherent it is because of this double function of violence, depending on the period in question. At one time Fanon is saying that the bourgeoisie "should not be combated" because of the risks to the overall development and he alludes to its self-implosion (WotE, 119–20). But he immediately states that "barring the way" is the only way (WotE, 121). "A persistent battle has to be waged to prevent the part from becoming a compliant instrument" (WotE, 126). Other times he seems to shy away from explicitly saying that violence is not the solution. It is the case when he calls resorting to violence a "shortcut, which dangerously combines spontaneity, simplistic syncretism, and little intellectual elaboration." It is an "abdication from responsibility" (WotE, 139) as if equating violence with irresponsibility.

Elsewhere he identifies the historic mission of everyone who wants to rid of colonialism to "authorize every attack aborted or drowned in blood" (WotE, 146). But he asserts this in reference to the "first phase of national struggle." Additionally, authorising attacks does not equal perpetrating them – he has never called to it. The struggle, the acts of violence come naturally from the masses but the details – how that happens and under which leadership – remain ephemeral.

Arendt is also a philosopher of commitment – she avoids ideologies but in fact proposes a grand project that has similar ideological consequences to Fanon's new humanism. Thinking and action are for her man-specific tools of existence. Arendt shies away from saying how politics should be done – instead her philosophy of responsibility concentrates on education. She saw first-hand how universities became culprits in the spread of totalitarianism. Herself an educator, she was most intensely interested in university life, in teaching and student movements. She was also committed to her work as a teacher – just as Fanon was to his profession of a medical doctor. She sees an educator in Fanon's texts – a more potent educator than she would be herself, because Fanon pushes students to action.

Commitment to being the other

The crucial epiphenomenon of Fanon's thoughts on violence is the ethical position of every one of those who do not pertain to the colonised, including himself. The only way to side with the colonised is to be them, to be unconditionally on their side, regardless of the urge to problematise and absurdly rationalise every choice and, as a result, stand in the middle – such inclinations are all Western and colonial. This is why, being Martinican, French and black, Fanon becomes an Algerian fighter. No middle way is left here if a new order is to be created – everyone responsible, everyone who understands and sees social relations clearly must take sides.

Arendt seems to be of similar opinion. Ethical systems, among them the system of human rights in particular, cannot be the answer. "The world found nothing sacred in the abstract nakedness of being human" (OoT, 299). Having one's own nation and state, this is what realistically makes human – a stateless person, a Jew in Nazi Germany, a Syrian in Hungary in 2015, a Nigerian in Calais in 2016, a Filipino in Saudi Arabia is not human, deserves no rights. The revival of traditional polity, in which all are equal, is her remedy. The random act of natality, the inequalities between individuals outside of the political realm make us unfree. Only in political institutions equality finally takes shape. Yet

Western humanity has diverted from the Greek understanding of a polity. Loneliness, rather than no-rule all-equal community is the experience of masses of people. She distinguishes between uprootedness (having no recognised place in the world) vs. superfluousness (being as if not in the world at all) (OoT, 474–75). The colonised, the refugees, Muslims, Arabs swerve between uprootedness and superfluousness.

> It is the same story all over the world, repeated again and again. In Europe the Nazis confiscated our property; but (...) in Paris we could not leave our homes after eight o'clock because we were Jews, but in Los Angeles we are restricted because we are "enemy aliens." Our identity is changed so frequently that nobody can find out who we actually are.
>
> (WR, 115–16)

Yet she is critical of pity. Who pities, keeps the power, instilling the inequality between the pitying and the pitied. Imagine to be in the place of another is what she is teaching, imagine yourself in an uncomfortable place – the place of the refugee, the place of the colonised, the place of the hated. She, too, is committed to being the other. Ourselves we are being the other through dialogue with ourselves, in which we develop moral imagination. The constant conversation in our heads, the hesitation, the pros and cons she takes from Socrates and Heidegger – this inner dialogue creates the right conditions for understanding others and for moral judgement. Through being the other we become moral.

Responsibility of the political

Arendt develops a unique understanding of responsibility by linking it with the political sphere, as Herzog explains, thereby removing it from the realm of morality.[14] Although intellectual's responsibility is different from the one of political actors an individual is responsible by being part of a community:

> Linking responsibility to belonging immediately delegalises responsibility. 25 years after 'We Refugees', Arendt still argued that one is responsible not because one acts under a predetermined law but only because one belongs to a group that acts, or has acted, independently of him/herself.[15]

Such a concept of responsibility – as being in fellowship with others and acting in the political sphere – is Fanon's revolution. For Fanon the responsibility is not only (perhaps not predominantly) personal – it rests on the state and political parties. Fanon's unarticulated definition of a political party builds on the same understanding of power that Arendt adopts. Because power comes from the opinions of people, a political party is a network that hears in what the villagers and city dwellers alike are pronouncing. But hearing this does not make a political party incapable of independent action. On the contrary, political parties must cope with social underdevelopment. Fanon's social project rests on ascetic and laborious living. The large youth bulge in African countries, prone to the dangers of the Western culture: pornography and, most importantly, alcohol, needs to be taken care of by the politicians. Families are too weak to balance that out. Western youth can resist the dangers of Western culture because there the intellectual societal development is even and the school, the family and the level of living are all strong bulwarks against these dangers (WotE, 136). Fanon would want the young to be interested in agriculture and education, rather than stadiums. In his attitude to sport it is interesting that he differentiates between the body and spirit-building function on the one hand and the hierarchy imposing and mind-boggling functions on the other. He is against producing professional sportsmen because sport becomes an industry and creates new heroes/leaders. Sport should rather be part of people's lives and thus empower them physically and mentally.

Although rigorous Fanon's prescriptions have not become outdated. The youth bulge in the Arab and African world, where more than 60% of the population is less than 25 years of age as of 2019, is the sole most challenging social group to satisfy. They all need or will need jobs in the future and they are all prone to strong ideologies that can tilt their development in unpredictable, progressless directions.

As noted by Alessandrini in the context of Arab/African revolutions of 2011 Fanon's "legacy of unsparing intellectual and political commitment" presents itself as still relevant, which is both a negative and positive ascertainment.[16] The positive side of it is that 2011 as a struggle against the authoritarian regimes in the Middle East, the remains of the colonial system in the exploitation and hypocrisy part, is a struggle that Fanon called for and it is in the making. The negative one is that little has changed in the 50 years that separate WotE and the Arab Spring.

There is more to Fanon's responsibility: it goes beyond the colonised – it is a responsibility for the humanity, both in the south and the north. The responsible and committed must work to change international relations too. "The nuclear arms race must be stopped, and the underdeveloped regions must receive generous investments and technical aid. The fate of the world depends on the response given to this question" (WotE, 61). More importantly still, it is a responsibility for bringing the south and the north together. It can be seen in his pre-WotE writings and lectures:

> The end of race prejudice begins with a sudden incomprehension. The occupant's spasmed and rigid culture, now liberated, opens at last to the culture of people who have really become brothers. The two cultures can face each other, enrich each other. In conclusion, universality resides in this decision to recognize and accept the reciprocal relativism of different cultures, once the colonial status is irreversibly excluded.[17]

Behind Fanon's philosophy of commitment and his preoccupation with the liberation of the south lies the most profound goal for the humanity: a reconciliation and unity between the south and the north.[18] Something Europe did not do, even though it had often come to the understanding of such necessity and proclaimed it in the essence of its thought. On the basis of this feature it has been argued that Fanon preached "radical empathy," understood as politics of solidarity and recognition with communities beyond one's own experience,[19] or on a methodological level, Fanon did not, as Edward Said claimed, borrow or adopt the European thought to colonial studies, but he "affiliated" it in the most profound meaning of the word, building an intellectual and moral "community of a remarkable kind."[20]

Political violence in international relations, on the international arena and colonial violence are closely linked, there's a "complicit correlation," a "homogeneity" (WotE, 40). "The colonized is today a political creature in the most global sense of the term." Fanon's universality of politics means its inclusiveness to all peoples of the globe – for Arendt too, politics must be a universal possibility since it distinguishes a human being. That universality can be seen in her take on responsibility and otherness. Arendt, as much as Fanon, can be read as responsible for the humanity, here in the words of Herzog:

> I am responsible when my free doing stands for other others (…) My responsibility fills the gap between my community and the world.[21]

Arendt objects to Fanon's commitment because her own commitment is to the fight against grand ideologies, especially violent ones. His philosophy is openly ideological. Her main objection to Fanon is that violence will not lead to new humanism because it leads to the death of politics.[22] The distinction between politics and war is necessary for there to be politics at all.[23] War is not politics – according to Arendt, politics is only possible in peace times. The political sphere is where human is human and not animal – it differentiates the human species from all other. The sphere of appearances as well as language are only attainable to people. Violence, however, does not speak, it is mute. Only words, not violence, can bring out new meanings. And, likely, Fanon would not disagree even though he, and Arendt too, would particularly emphasise the function of praxis. Fanon uses words to bring out new meanings, just like Arendt suggests, still being committed to praxis and the action of steering the nation in the right direction. He is about doing; Arendt is about thinking. She does not believe in "correct directions" and is weary of ideologies. He commits his life and family to the cause – she commits it to thinking – she sees the origins of doing in thinking, in the dialogue with oneself, which is a gate to being able to dialogue with others. But thinking is not easy and certainly not equally easy in all circumstances. In HC Arendt admits that "it is in fact far easier to act under conditions of tyranny than it is to think" (HC, 324).

Arendt finds that "of all equalizers, death seems to be the most potent," when it does play a political role, which it does rarely. In war/battlefield the noblest, most selfless deeds occur daily – only there the proximity of death intensifies human vitality, one's own death serves immortality of the group. Yet death is "the most anti-political experience there is" (OV, 67). Elsewhere, however, she notes that death, out of all equalisers is the most just one – and so it can be the mother of a new world (OoT, 328).

> Society equalizes under all circumstances, and the victory of equality in the modern world is only the political and legal recognition of the fact that society has conquered the public realm.
>
> (HC, 41)

As Habermas explains, Arendt's concept of "radical equality" rests on language: the ability to speak and communicate, which, in turn, instils unimpaired intersubjectivity.[24] But her concept, like Fanon's, is never tested against the realities of the times:

> Power is actualized only where word and deed have not parted company, where words are not empty and deeds not brutal, where

> words are not used to veil intentions but to disclose realities, and deeds are not used to violate and destroy but to establish relations and create new realities.
>
> (HC, 200)

The question that poses itself here is which – Fanon's new man or Arendt's words with true meaning that create power – is the more elusive and disrespectful of reality? Are they not both similarly extraordinary even if worded in diametrically different registers of language? They both call for the same true representation of reality in words and new relations in deeds.

Fanon's praise of violence rests on equating it to being fully alive as opposed to being quiet and dead. The proximity of death stirs vitality (Arendt found that function of death and violence to be true only in war). Sorel's idea was similar and so was Spengler's and Pareto's who argued that bourgeoisie had lost its energy in the class struggle and only the awakening proletariat could save the European bourgeoisie by reinvigorating its fighting spirit. Sorel saw the bourgeois as peaceful, bent on pleasure, hypocritical and deprived of the will to power. The worker was the only creative element of society: a producer of new moral qualities, which turn out to be honour, desire for glory, fighting spirit ridden of hatred. The worker features all the qualities that the bourgeois once had before he became hypocritical. Upon discovering the Establishment or the System the worker turns to violence. But Arendt's objections to Sorel or Pareto cannot be valid for Fanon, who – even if strongly influenced by Sartre – eschews the bourgeoisie-proletariat division. Fanon's new man is romantic – he's a southern farmer.

In the peasantry Fanon finds spontaneity, the appreciation of which connects him with Arendt. Both seem fascinated with unplanned joint action of people – a fascination that perhaps awakens naturally when observing an impromptu cooperation of strangers toward a goal. Who would disagree that it uplifts the spirit? In the French, American and Russian revolutions Arendt marvels at the "spontaneous power organs," spontaneously formed clubs and societies that have little to do with municipal bodies, the National Assembly, revolutionary parties or trade unions (OR, 240–59). Spontaneity allows the people to discover action, something they did not know it existed (OR, 246). Spontaneity is an elementary condition of action, which – through the creation of a network of councils – begins to constitute a collective entity, a state (OR, 267). As Gines skilfully noted there is inconsistency in Arendt's objecting to the unpredictability of violence in OV (presumably advocated for by Sartre and Fanon) and in her appreciation of spontaneity, which is a characteristic of action as opposed to labour and work (HC).[25] In WotE's Chapter 2, entitled "Grandeur and Weakness of Spontaneity," Fanon strikes similar tones to Arendt: "the programme of every spontaneously formed group is liberation" (WotE, 83) and spontaneity inherently belongs with the revolutionary peasantry. But he warns about the weakness of spontaneity: it is creative in the revolutionary moment but useless as a doctrine (WotE, 85), and it can be artificially created. An orchestrated plan can deceive by having a semblance of spontaneity. (WotE, 87). Yet, taking no notice of Fanon's take on spontaneity, in OV Arendt forcefully rejects the possibility of any creativity or vitality in anti-colonial violence, even though herself she had been on occasion of different opinion, as Hanssen noted:

> Arendt allowed for revolutionary violence, as long as it remained spontaneous, ephemeral and got to be replaced by a higher order in which politics reigned supreme.[26]

Instead Arendt reminds us that in pre-philosophic political thought, since death was the equaliser, politics was the sphere of immortality that assured deathlessness. To Hobbes it was not death itself that equalised but the fear of everyone being able to kill anyone else that leads to a political agreement (OV, 68). The experience of strong brotherly emotions that collective violence rouses misled Fanon into believing that a "new man" can be thus created. That kind of emotion is extremely transient, occurring only in danger.

However, Arendt is also charmed by the possibility of man recreating himself. The allure of Lawrence of Arabia is an example in point. Arendt sees him as one who cannot stand the hypocrisy of the British society of the time and endeavours to recreate himself in Arabia, the true self he finds in partaking in the winds of history (OoT, 218–20). He is the epitome of the post-war elite who yearns for "*losing their selves* and the violent disgust with all existing standards, with every power that be" (OoT, 327).

She would claim that Fanon used Sorel's categories even though his far richer than Sorel's experience with violence "spoke clearly against them" (OV, 70) – in the footnote she supports this claim by citing Barbara Deming, a human rights activist, that – in her opinion – Fanon might as well plead nonviolence. Deming proposes an exercise: let us substitute in WotE the word "violence" with "radical and uncompromising action."[27] In effect we would get a text that praises nonviolence. In that same essay Deming uncovers a Fanon that is different from the one Arendt reads: Deming sees a revolutionary who calls for invention, concerted action and new humanity, a writer that is misunderstood as calling to arms.

Arendt sees the social revolution of her times – the apotheosis of life and love-making – as a response to the actuality of a doomsday in nuclear war (OV, 73), which makes people ponder about the visceral biologism of life. Arendt believes that the philosophies of Bergson and Nietzsche are being revived in the Sorelian form, because they rest on biological interpretation that growth is immanent to power – power either grows or dies, like all life forms in nature. Fanon seems to subscribe to this interpretation. Such political organicism Arendt perceives as the most dangerous one theoretically (OV, 74). Violence is justified on the ground of creativity – death and life, destruction and creation are merely two sides of the same natural process. In this line of thinking we get to the absurd thesis that the more we die, the more we live, which she attributes to Fanon and faults him with political organicism. Arendt goes further to claim that such biological metaphors resonate particularly well if the matter has to do with relations between races. We know biological processes first hand, they make sense and seem cohesive as a system. Biological analogies have the power of turning prejudice into ideology – systematising the prejudice.

But Fanon, like Arendt, is against these biologisms. "Even in acute periods of armed struggle, biological justifications are never used"[28] Fanon declared long before writing WotE. It is true that in WotE the colonised are driven by their guilt complex ("do I even want to live?"), which makes them feel inferior. Guilt, Fanon says, drives violence. Violence as an unavoidable, natural outcome of the guilt complex is a tragedy that cannot be avoided. Apart from the psychological and psychiatric stance that Fanon naturally adopts he shares Arendt's opposition to biological explanations of violence, which he sees as colonial. He notes high Algerian criminality before 1954 – one of the highest in the world, the colonists would claim. The Algerian is seen as a habitual, savage and senseless killer, capable of killing over a gesture or allusion. The connection between Muslim psyche and blood, the animal and human throat slitting so that blood oozes still reverberates today and is reminiscent of Jewish blood libels. The North African loves extremes and is violent (WotE, 223). Melancholic Algerians do not try to commit suicides, as is often the case with non-Algerian melancholics, but they try to kill others: it is their biological disposition. Fanon gives examples of two "experts" who dehumanise Algerians and Africans in biological, "scientific" terms: (1) Professor A. Porot, who is of the opinion that the North African does not have a cortex, and (2) Dr J.C. Carothers, the author of "The African Mind in Health and Disease: A Study in Ethnopsychiatry," issued by the World Health Organization in 1954, in which he conceives of the African as a lobotomised European. According to them, it is the brain structure that is responsible for the native's inaptitude and animal impulsiveness. Fanon uses the colonial context to explain criminality. An Algerian turns on another Algerian because they physically meet daily, they are all exhausted with inhumane conditions and turn their aggressiveness on each other. In yet another analogy to the history of Jews as "others" Fanon says: "Any colony tends to become one vast farmyard, one vast concentration camp where the only law is that of the knife" (WotE, 231).

Arendt is also against biological explanations of violence. She discredits scientific research on aggressiveness in animals that has claims to results for humans. She particularly objects to the effect of justifying violent behaviour on the basis of social and natural sciences research. Arendt laughs off aggressiveness as a natural instinct (similar to the nutritive and sexual instincts) that, when suppressed, causes frustration. In biologisms the concept of man is positivistic: the human being differs from animals in that it reasons and so it is able to create science. Science discovers laws and teaches the human being how to suppress what is irrational, animal-like in us – like violence, which is irrational. Arendt disagrees: violence is neither beastly nor primarily irrational (OV, 62).

Man creating himself in the public space

Fanon and Arendt agree that man creates himself in the public space, in the political sphere. To Fanon the colonial man is a creation of the metropolis. What can be done to change that? One can either try to build oneself on the basis of universal values or (re)discover one's own culture (tradition, heritage, history etc.) and build the new self on the basis of the rediscovery. That rediscovery, however, cannot be made. Values, which inadvertently turn to be Western – individualism, dignity, beauty, the heritage of the Enlightenment – are a mockery to the colonised. He does not know human dignity, because he has never experienced it – he experienced "indignity" under the name of "dignity," values turned to be un-values. As was noted, Arendt, too, finds some of those Western values, human rights in particular, void of meaning. The most important true value to the colonised is the land, which gives food and allows for survival.[29]

> Fanon was not enamored of this reduction of human existential horizons to the requirements of survival; but neither did he share

> the aristocratic contempt of a Hannah Arendt for a revolutionary impulse driven by the compelling needs of the "vital process," the illicit invasion of the "public realm" by the call of necessity.[30]

The colonised has neither been able to retake the universality of values from the colonist, nor has he had time yet to build and ascertain values anew. The only heritage of the colonised is a terrifying void. "We are all in the process of dirtying our hands in the quagmire of our soil and the terrifying void of our minds" (WotE, 139). The old culture is gone, the new one is absent, the only one now present is the Western culture, in which the colonial intellectuals are submerged. When they want to break with the Western culture and connect with the old national culture of their own they are "terrified by the void, the mindlessness and the savagery" (WotE, 157). Self-denial becomes a necessary element to recreate oneself.

Taking Merleau-Ponty's phenomenological stance Fanon sees the body as a form of consciousness but also discovers that stance's limitations in the Third World. If the man of colour was to take the world through his lived experience – an experience of oppression, exclusion or other inferiority – he would never escape these injustices, nor would he be able to create. Simultaneously Merleau-Ponty's phenomenology gives Fanon a way out of this viscous circle: the body as an actor. The praxis of the body will lead to the creation of completely new values. In the struggle man creates himself physically and normatively: in the body and in the mind.

Violence or struggle is the only concerted action that the colonised are capable of. There is no alternative action that would unite them, there is no public space. When there are no values and a void of the mind, the body in motion against the colonist is the only available action. Nonviolence struggle does not make sense because it requires values that are non-existent and the only tool of the colonised – the body – is inactive or at least less active in it.

In this context Fanon explicitly disregards religious (also jihadist) incentives to struggle: "the Algerian fighter has an unusual way of fighting and dying, and no reference to Islam or Paradise can explain this spirit of self-sacrifice when it comes to protecting his people or shielding his comrades" (WotE, 221). Instead, he offers a messianic notion of self-sacrifice for humanity, which leads Fanon to conclude that the struggle of the colonised will create a new humankind on the ruins of the decayed modern culture.

In colonial countries only the peasantry is revolutionary, Fanon observes, and so the peasant must create the new man. Peasants defend tradition, they remain disciplined, even if a peasant society "sporadically generates episodes of religious fanaticism and tribal warfare" (WotE, 67). This quote is yet another argument against the claims that modern jihadism was the kind of struggle that Fanon gave rise to. He may have overestimated the rarity, with which fundamentalism surfaces but the processes that strengthened these ideologies came later, in 1970s and 1980s.

The choice of the peasant for the ideal revolutionary reveals Fanon's difficulty in striking a balance between the distrust of the intellectual on the one hand and the tribal/religious on the other. The farmer escapes both traps: he cannot live in mythical reality because he's organically tied to the land, yet he is non-elitist, traditional and disciplined. The choice of the peasant is Fanon's romantic quest for the pure southerner – a pure fighter who creates himself as a man, a partner of the northerner. Leo Zeilig rightly picked out a recollection of Claude Lanzmann's meeting with Fanon, when he describes the FLN fighters as "polymaths who not only fought the French by force of arms with utter purity and self-denial, but also studied philosophy" – Fanon was looking for a peasant-warrior-philosopher,[31] a pure hermit who – through the only available action to him as a non-European, that is struggle or violence – makes himself a human, a thinker, a philosopher. Such a man will not allow the Third World to fall into the Dark Ages that Europe had fallen. The new man will come from the colonised as a gift for the humanity, both in the south and the north.

Like Arendt, Fanon is critical of consumerism and materialism. His remedy is education for the masses (WotE, 137) and his aim – politicisation of the masses. Through a new concept of man, a concept about the future of mankind, the non-existent public space in the Third World will be created (WotE, 142). His project is immense, provoking Arendt to read Fanon as a dangerous dreamer but in fact their calls are not dissimilar.

Arendt's attitude towards the creative aspects of struggle seems uneven. She criticises the concept of creation through violence in OV but in her other writings she puts forward that very concept of creation through violent political opposition, in accordance with Fanon. In OV she leaves no doubt that she finds Fanon's idea for the creation of a new man in struggle – a false one. Her first argument against Fanon's assumption that anti-colonial violence will lead to a freer new global order led by the once oppressed is that he mistakenly assimilates violence with historical development. Her second argument, but connected with the first, went against the "creative madness" of violence, for there is no creativity in violence. Arendt admits that the very basis of leftist humanism is the idea of man creating himself: in Hegelianism man creates himself through thought, in Marxism – through labour. Thinking and working, however, are not violent actions, do not entail killing. Violence may seem to be the sole disruptor of processes in history. But it is not violence, Arendt claims, it is *action* that disrupts otherwise predictable processes and mere behaviour (OV, 30). These thoughts coherently correspond with what she lays out in HC.

She, nonetheless, takes the side of the violent student rebellion of 1968 on the grounds that (1) their rebellion was against powerless bureaucratic governments and (2) they stopped short of taking over the power. They withdrew peacefully once their postulates were met. The student rebellion had moral roots and causes – the rebels have proven that quiet manipulation (television, advertising) cannot be successful in a free society despite an upbringing in the culture of selfishness (OV, 28). They revolted against science, the positivist progress that does not result in the progress of mankind. She calls the then system of government a rule by Nobody – because it is a rule of bureaucracy, a rule where the "enemy," the genuinely responsible cannot be identified (OV, 38). It is no longer an absolute rule, neither a democracy, nor an autocracy. She implies that the political systems in the West – "rule by nobody" – invite violence because responsibility becomes dispersed. Under these circumstances violence is justified.

Action is a feature that determines who has power, and – in between the lines – Arendt calls for action. When a minority disrupts a lecture, it can do so because the majority allows it to, does not oppose it actively. A passive majority becomes an ally of the active minority. Minority can have greater power than the mere headcount. "The merely onlooking majority, amused by the spectacle of a shouting match between student and professor, is in fact already the latent ally of the minority" (OV, 42). People revolted with very unfavourable odds, being disproportionally weak in comparison to the might of the state. But state superiority only lasts as long as its power is intact, meaning as long as commands are obeyed. When people disobey the system, it changes abruptly. Where commands are not obeyed, means of violence are irrelevant, only opinion and numbers who share it matter. "Everything depends on the power behind violence" (OV, 49). Therefore, Fanon's description of the realities in Algeria in the 1950s corresponds well with Arendt's revolutionary moment: when people (Algerians) disobey the system (French), they begin to fight it.

Weak governments last because no one tests them to reveal their weakness, or because they are lucky not to be engaged in a war that reveals it. Following this logic, the French rule in Algeria continued because either it had not been tested or it was powerful, meaning had vast support. Both might have been true for Algeria at different points in time. What is missing in Arendt's argumentation is the definition of a test or war that reveals state's weakness: are they usually not outbursts of violence, sometimes prolonged if the disproportion of strength is to the structure's favour, rather than or alongside non-violent struggles?

Analogically to Fanon's argument for the politicisation of the south, Arendt seems to be making the case for a repopulation of the public sphere in the north. The enemy of the political sphere is bureaucratisation. The greater the bureaucratisation the more attractive violence seems, because there is no one left to argue with. Bureaucracy deprives of political freedom and the power to act – "when all are equally powerless we have a tyranny without a tyrant" (OV, 81). With the shrinkage of the public realm, big party machines – both in the East and the West people have been deprived of political action, albeit differently. Man is a political being because he can get together with others and act in concert to reach goals that he would have never even thought about if it wasn't for his human condition for action. Free citizen is a citizen-co-ruler (she cites Pavel Kohout, a Czech novelist) in participatory democracy – we are in the need of a new example, Arendt warns, if we are not to become chicken or rats ruled by an elite who thinks that people in think tanks are thinkers and computers can think. The allure of violence grows because, indeed, it gives a semblance of collective action (OV, 82). People are naturally hungry for action – if they cannot act politically, they are kindled by violence. It may partly explain why terrorism and violence prevail in the Middle East where the public sphere is still limited and politics in the West too becomes more and more violent, or at least is brutalised and less and less moral. In fact Arendt's insights mimic Fanon's argumentation that the colonised is a political creature in the most global sense of the term. The colonised that is creating its public sphere for the first time is becoming a free citizen, a citizen-co-ruler finally. Fanon would perhaps agree that if the colonised cannot act politically – and they cannot – they are kindled by violence. If violence is the only action possible, regrettably, then action needs to be executed through the only mean possible. As we shall see that way out is plausible also for Arendt.

Arendt called politics the space of appearances (HC, 198), where the human has a political mask in this appearance. Hypocrisy and violence are both dangerous in the sphere of appearances. Hypocrisy means putting a mask on a mask. Words lose meaning, responsibility is dissolved. Violence tears all masks, and the mere human trunk remains; violence also tears off the necessary political mask. Fanon, with his claim of being both the slave of his appearance and the slave of the coloniser stretches Arendt's concept of politics twofold. Firstly, by showing that there is a mask that never comes off: black skin. And secondly that nonviolence or value-based revolutionary student movements are impossible in the Third World because of the nonexistence of public space. Fanon's concept is about tearing off the masks that have grown into the skin – it's a violent and painful act. In *Between Past and Present* Arendt herself gives violence a public space, creating value when discussing resistance under Nazi occupation:

> they had discovered that he who "joined the resistance, found himself" … that he could afford "to go naked". (…) they had been visited for the first time in their lives by an apparition of freedom, not because they acted against tyranny (…) but because they had become "challengers," had taken the initiative upon themselves and therefore, without knowing or even noticing it, had begun to create that public space between themselves where freedom could appear.
>
> (BPF, 4)

For Arendt too, it seems, public space and freedom is the fruit of the struggle against occupation. As Serequeberhan put it "the colonized non-European, just like the generation of Europeans who lived under Nazi occupation, finds freedom and liberation in the resolved confrontation with the colonial apparatus."[32] Yet Arendt refuses to admit it in OV. Why would she argue with Fanon? Were colonised not under occupation? Was their misery qualitatively lesser than that of the occupied French or Polish? Was it because of Eurocentrism or ignorance, asks Serequeberhan, that she fails to see European and non-European struggles alike?

The fundamental difference and problem that makes Arendt's ultimate agreement with Fanon impossible is that her concept of politics is completely deprived of socio-economic relations. Habermas noted that such an understanding of power and politics makes Arendt ignore structural violence,[33] which is the method of colonialism altogether. If the basis of power is the contract between free and equal parties, then the colonised cannot be producers of power: they are neither free nor equal.

The questions about Arendt's stance on power relations in the south are all the more pertinent that she, in her other writings, is seemingly in agreement with Fanon. She thought, for example, pacifism was unrealistic (OoT, 441–42). She advocated for the establishment of a Jewish army. Such a construction would constitute the beginning of Jewish politics, she claimed.[34] Violence could then be creative. She faults Fanon with using rhetoric she herself praised. "Fanon's worst rhetorical excesses, such as *hunger with dignity is preferable to bread eaten in slavery*" (OV, 19): we all know it is not true – she says – for we know the processes in our body. But for Fanon this is the way to create the public sphere and discuss values that have a meaning. Exactly the postulate of Arendt herself when she talks of words losing their meaning. Arendt argued that politics was never for the sake of life (HC, 37) and life was not the highest good. Dignified hunger, a concept similar to that of the Jewish army, was precisely the way to the creation of the immortal and the free – the public sphere.

> Freedom is actually the reason why men live together in political organization at all. Without it, political live as such would be meaningless. The *raison d'etre* of politics is freedom, and its field of experience is action. (…) In order to be free man must have liberated himself from the necessities of life. … Freedom needed, in addition to liberation, the company of other men, who were in the same state and it needed a common public sphere to meet them – a politically organized world, in other words, into which each of the free men could insert himself by word and deed. (…) Freedom as a demonstrable fact and politics coincide and are related to each other like two sides of the same matter.
>
> (BPF, 145–47)

Here Arendt comes very close to Fanon, because she confirms his diagnosis that only true liberation of the people of the south will make them a polity. Arendt explicitly excludes family households, tribes, clans as well as despotic and tyrannical systems from political entities: there the public sphere cannot exist because freedom is absent. Actions in these unfree organisations are ruled not by freedom but by the necessities of life and its preservation. Life in colonial and neocolonial realities of the Middle East is precisely ruled by the necessities of life and its preservation. "To *be* free and to act are the same" (BPF, 151) – she talks not of a freedom of will but freedom to create something that was not there before. Creation through action is dangerous for life loses its primacy in the public sphere, which is the realm of the immortal. Fanon's struggle, a violent action with disregard for life is to create something that was not there before – a political sphere and a new free man – without really the intention and aim on the part of the people who are acting to create it.

Political nation through bonds

Both Arendt and Fanon see the resilience of a society in the bondage that people create between each other and both are wary of nationalism. Arendt's fist fascination with the United States was exactly with how connected people were on the grassroots level. She articulated the *spatial* character of public life – it happens in space, in the public space between the people. This public space was non-existent for the colonised in Fanon's Algeria. The public space of the colonist was a realm for its own sake, a prolonged public sphere of France, but the colonised did not even form an underground public space for their own citizenship to flourish – being unfree they could not. The struggle for freedom finally sowed the seeds of a public sphere in North Africa but whom would it be for? Who forms the new community – is it a nation, what is its culture?

Fanonian nation is a form of consciousness based on a shared fate – for Arendt, too, freedom needed liberation and the company of other men who shared fate. Along that shared fate – the fight for freedom – re-enlightenment of the people and oneself takes place on the "full measure of men" (WotE, 218), on what a man is capable of doing to another man and on how people need to reconnect. The struggle facilitates re-establishing one's weight as a human being. "If culture is the expression of the national consciousness, (…) national consciousness is the highest form of culture." The greatest possible cultural manifestation is a "conscious, organized struggle undertaken by a colonized people in order to restore national sovereignty" (WotE, 178). Once this struggle is over, there is no longer colonialism nor colonised, a new humanity is born. Struggle creates this new humanity under certain conditions: it needs to mobilise every societal level, express the expectations of the people and stem from complete cooperation between men. Arendt admits that collective violence creates a strong although short-lived collective bondage. In the short term it is stronger than individual bonds but in the long run it fades as opposed to still strong individual ties. Arendt writes, in reference to Fanon, about an "intoxicating spell of the practice of violence which binds men together" forming the great organism of violence (OV, 67).

Fanon prioritises the national stage of the struggle over larger international unions. He does espouse cosmopolitism but in a specific way. National consciousness ("which is not nationalism," Fanon emphasises, WotE, 179) is international in the meaning that it is continent-wide, relates to African consciousness as a whole. An African who is responsible to his national culture is also responsible to African culture. Inbuilt in the national consciousness is international consciousness.[35] He believes in the strengthening function of solidarity of non-colonial peoples. That solidarity is alas absent. Internal divisions today of the post-colonial world add to the validity of the need for continued struggle. Even the Palestinian case, which seemed to be the only political issue that would unite the Arab/African world is used instrumentally by many Arab governments. Arendt would claim in OV that it was wrong to think that there was a "unity of the Third World" – Third World is not a reality but an ideology (WotE, 21). Fanon would most likely agree – the Third World existed not as an objective geographical fact but as a conglomerate of peoples who had been colonised and subordinate to Europe.

Fanon demonstrates interesting attitude to nationalism. Nationalism is not a programme or a doctrine as an end in itself. Switching from national consciousness to political consciousness is a necessity (WotE, 142), because only then will a nation be built. National consciousness is not the aim in itself but a means to political consciousness. He admits the advantages of political nationalism and turns against the ethnic one, which he calls chauvinism. "There is a constant pendulum motion between African unity, which sinks deeper and deeper into oblivion, and a depressing return to the most heinous and virulent type of chauvinism." (WotE, 104) There is only one step between chauvinism and tribalism.

> National independence brings to light multiple realities ... the people who in the early days of the struggle had adopted the primitive Manichaeanism of the colonizer – Black versus white, Arab versus Infidel – realize on route that some blacks can be whiter than the whites.
>
> (WotE, 93)

He calls them "profiteering elements." The chaotic time immediately after independence gives birth to the betrayal of the elite. Militants realise that they are unwittingly building another system of oppression, an indigenous one. And vice versa: "Some members of the colonialist population prove to be closer, infinitely closer, to the nationalistic struggle than certain native sons" (WotE, 94). André Mandouze, for example, co-founded *Témoignage chrétien,* Christian resistance that i.a. first saved Jewish children and later in Algeria was the *spiritus movens* behind the AJAAS, the Algerian Youth Association for Social Action. Mandouze supported the Algerian revolution and was later a go-between the FLN and the French government.[36] Mendouze is an example of an ethnically non-Algerian member of the Algerian political nation.

The bourgeoisie, local or foreign, has been unable to "enlighten" people in general and about the notion of the nation specifically. The young nations are still prone to internal divisions that flourish under the slogan "replace the foreigners" (WotE, 105). Since it is repeated by the elite, the people use it on lower levels of social strata and call for expulsion of anyone of their fellowmen, who has been identified as a stranger/foreigner – of different tribe, ethnicity, religion, sect. It is not only the foolishness of the elite's policies but also the remains of colonialism in the form of privileged provinces/areas that enables the return to tribalism and racism. In Fanon's own example in Lebanon, Arab and Muslim imperialism is blamed for colonialism altogether; elsewhere the opposite is true: the Christians are vilified as enemies of national independence. To Fanon religious tension can be a form of racism, especially that Africa is racially divided into the white north and black Sub-Saharan part. The elite, "who had assimilated to the core the most despicable aspects of the colonial mentality take over from the Europeans and lay the foundations for a racist philosophy" (WotE, 108). White Africans (Arabs) will call black Africans "niggers." Fanon saw then what today too can be vividly noticed: for example, the racism of many Libyans towards Sub-Saharan migrants working in Libya or being enslaved by the people smuggling industry.[37] The popular word for a black person in Arabic is *'abd*, which means "slave." Western bourgeoisie is offensively racist (it believes in equality; therefore it requires that the inferior rises up to the level of the superior), while the non-Western racism is defensive, established on fear (WotE, 109).

The new national consciousness, born in the newly independent colonies, is fragile. It is easy to switch back from nation to ethnic group, from state to tribe – a regression to ethnic nationalism and tribalism is detrimental to national unity (WotE, 97). Weak national consciousness, narrow-minded nationalism, is caused not only by oppression but also by the local bourgeoisie's mediocrity. That group is made of business elite and university graduates who are traders, land owners and professionals but they are (so much) inferior compared to the metropolis elite. They are not "geared to production, invention, creation, work" (WotE, 97). The only way this bourgeoisie connects with the local is by its cult of local products, detrimental to local economies. This forces the underdeveloped countries to continue shipping raw material, grow food for Europe etc. "The national bourgeoisie replaces the former European settlers as doctors, lawyers, tradesmen, agents, dealers" (WotE, 100). They are a conveyer belt for capitalism (a mask for neocolonialism). They mimic Western bourgeoisie in its decadent aspects without having accomplished the initial phases of exploration and invention, which have made the Western bourgeoisie flourish. Such superficial mimicry of Western bourgeoisie kills political nationalism. The Western elite helps in this endeavour by expecting from the south nothing more than exotism, casinos and hunting grounds – Fanon notes. Indeed, tourism was only opening in the underdeveloped world for the West half a century ago but is in full swing today. The ex-colonised bourgeoisie becomes a "a party organizer" – the country becomes a "bordello for Europe" (WotE, 101).

Clearly the non-xenophobic nationalism that Fanon is looking for is a political creation of the politicians and the elite: they are responsible for steering people's opinions in the right, communal, yet anti-racist and anti-tribal direction. Instead, however, in the first period of independence the elite copies the colonist, disregarding its mission to educate the people about what nation is. What happens is merely that posts previously held by the colonists are taken by the colonised bourgeoisie who do not have the welfare and progress of the nation as their goal. Science too, devoid of any humanist element, merely repeats European works on ethics and political philosophy (WotE, 109). Incompetent politicians instil bourgeois dictatorship with a one-party system oppressing the people.[38] European political solutions are employed without the realisation of what caused these solutions to be used in Europe in the first place. Stagnation lurks when the power announces yet another megaproject sinking the economy into submission to loans and donations. These mega projects have become over the decades a means of temporary survival for governments that lose power. Authoritarian governments in post-colonial countries of the Middle East have used it too: the government in Egypt first irrigated the desert with the Toshka project[39] and is currently moving the capital from Cairo to a newly built city of the future, in Libya Gaddafi was famous for similarly grandiose ideas like the manmade river in the dessert,[40] in Saudi Arabia a futuristic multibillion dollar Neom city is being built in the dessert. To Fanon, through such projects, the leader, who once had the moral authority, becomes a "CEO of the company of profiteers" (WotE, 111). With a sense of shame and unwanted disillusionment he concedes that the once revered leader's discourse is now to pacify and mystify the people, tell them to fall in line. He does that by recalling the colonist times and arousing pride on the victory long gone over the foreigner. He is the head of the party, but there is no party system – the party only hands down the orders from above (WotE, 115, 125). "Favours abound, corruption triumphs, and morals decline" (WotE, 116). The former militant becomes an informer.

The liberator turns into the omnipresent security apparatus: the corrupt army and the Mukhabarat. Police get increasingly provocative and brutal.[41] It was the police brutality that directly led to the revolts of 2011 in Tunisia and Egypt. Seven months before the uprisings in Cairo, Khaled Mohamed Said was brutally killed by two policemen in Alexandria in bright daylight.[42] Morals dive low to the point that revelling in materialism spreads limitlessly – a feature of the wealthier part of some of the Arab societies that cannot escape the eye today. This corrupt new leadership is focused on Europe (or the US), spending money and time there. The difference further down the timeline is that Europe, the US or other wealthy Arab states of the Persian Gulf move to poorer Arab countries anew in the form of store chains, luxurious goods, secluded neighbourhoods. The army takes over from the party: more repression and corruption follows – more discontent grows in the people. Fanon's conclusion is that "the bourgeois phase" of liberation should only be allowed if it is capable of creating a national culture and reforming the economy. Having described this socio-political reality of the newly independent Third World, which dangerously still resembles that of today, Fanon warns that the result of such a semi-colonial state can be fascism (WotE, 116).

In Chapter IV, "On national culture," he makes a claim that discovering a national culture from before the colonial times affects the psycho-affective equilibrium of the colonised in a substantial, positive way (WotE, 148). It contradicts the colonists' claim that the purpose of colonialism was to save the people from darkness. Since colonialism did not think of the colonised as Angolans or Algerians but as inferior "niggers", its racism was uniforming and continental. Likewise, continental was the culture in response to this, the culture of "negritude": i.e. Negro literature rather than Angolan literature. In the case of Arabs the struggle for national liberation was linked to a "cultural phenomenon" – the awakening of Islam. Intellectuals have vehemently tried to revive the memories of the golden age of Islam in the 12th–14th centuries. Fanon hails the Arab League and Arab unity at the time, which in retrospect has proven to be farcical.[43] Arabs' cultural experience is not national but Arab – hence they have a more "global" perspective (WotE, 152). However, economically they are not independent but intrinsically connected to the Mediterranean. Politically Arab states are heterogeneous and incapable of cooperation. Supranational culture (Arab-Islamic) is cut off from reality – it smoulders emotions, is neither productive nor substantial (WotE, 154). National culture is necessary, otherwise the nation will become "a body of angels," rootless and colourless individuals.

In an article for *Presence Africaine* in 1959 Fanon writes of a nation that is the result of a "concerted action of the people" (*née de l'action concertée du peuple*), that changes the state, status and the future of the people – it can only exist in the forms of exceptional cultural fertility.[44] The nation here is created not on the basis of ethnicity but on the basis of an act of will, a joint action. It is the foundation of political nationalism as opposed to ethnic nationalism. Fanon warned that if "nationalism is not explained, enriched, and deepened, if it does not very quickly turn (...) into humanism, it leads to a dead end" (WotE, 143). Only humanism protects from tribalism.

Arendt supplements Fanon's thoughts on nations and nationalism with her take on Western nationalism. Hanssen notes that Arendt (and Edward Said) "shared Fanon's unease with the pitfalls of national consciousness."[45] According to Arendt, the traditional notion of nationalism decayed through imperialism. The more unable the big European nations were to integrate other nationals into their societies, the more oppressive they became. Eventually, only in the colony, outside of the metropolis was a Brit finally British, a Frenchman – French.

> The solution of the riddle was imperialism and the answer to the fateful question why the European comity of nations allowed this evil to spread until everything was destroyed, the good as well as the bad, is that all governments knew well enough that their countries were in a secret state of disintegration, that the body politic was being destroyed from within, and that they lived on borrowed time.[46]

Imperialism carried out the decline of the nation through chauvinism, which served as a bridge between the two and was a natural product of nationalism. Later Arendt found the bridge between the two to be precisely: tribal nationalism and outright racism (OoT, 152). Tribal nationalism – resting on a belief that "its own people are surrounded by a world of enemies" and being a "perversion of a religion which made God choose one nation" (OoT, 226, 242) – is a danger recognised by both, Arendt and Fanon. The reason for its growth in Europe at the turn of 19th and 20th centuries was rootlessness – having no defined home but feeling at home where other of the tribe were. Membership in a political community comes with citizenship and participation in the public sphere, not with ethnicity, religion or any other characteristic. In OV she calls in the West for what Fanon calls in the south: participatory democracy. It is the only positive slogan of the new movement that echoes in the East and West (OV, 22). It is directed against the "Western representative democracy" and Eastern one-party systems. Participatory democracy (see Table 2.2) is a system of a political community, a nation that has power, while terror is a form of government that is installed when violence has destroyed all power and remains in control (OV, 54). The more social cohesion, the more linkages between people, the less probable that power will give in to violence, then to terror.

Arendt's ending of OV is a fascinating self-debate about the future that – as readers in 21st century we see – has already arrived. She sees disintegration processes (in 1960s Europe was only beginning to unite) in public services, increasing road accidents, pollution etc., accompanied by the decline of party systems (transforming to the needs of mass populations). She already notices a growing, global resentment to bigness (which in today's terms may as well be globalisation) in the form of nationalism and a tilt to the right (OV, 84). Ethnic "nationalism" (Arendt uses quotation marks) is now a danger to well-established nation-states such as Great Britain and France. The United States is rushing in the same direction. She is against monopolisation of power in the form centralisation because it reduces authentic power sources, which always lie in the people. In her writings, most notably OR, Arendt would look for and be impressed by attempts in 18th century United States and France at direct democracy – the people congregating together so that they could speak and act jointly. In *Civil Disobedience*, which was a prelude to OV, Arendt convincingly defends that power.[47] Not only was it futile, she would argue, but also dangerous to give people constitutional citizens' rights and put the fate of the state in their hands without giving them the opportunity to be citizens in the first place. The ballot box and the election day is not enough. Fanon's interested in the same: in people's power in the villages, in the unseen political activity of the colonised, which will, in turn, make citizens of the former colonised.

Notes

1 Lindqvist makes a convincing case (just as Arendt did in OoT) that Holocaust was the result of colonialism. See Sven Lindqvist, *'Exterminate All the Brutes': One Man's Odyssey into the Heart of Darkness and the Origins of European Genocide*, trans. Joan Tate (New York: The New Press, 1997).
2 Joshua Cole, "Massacres and Their Historians: Recent Histories of State Violence in France and Algeria in the Twentieth Century," *French Politics, Culture & Society* 28, no. 1 (Spring 2010): 110.
3 See: Jean-Louis Planche, *Sétif 1945. Chronique d'un massacre annoncé* (Paris: Perrin, 2010).
4 Richard J. Golsan, "Memory's Bombes a Retardement: Maurice Papon, Crimes against Humanity, and 17 October 1961," *Journal of European Studies* 28, no. 1/2 (March 1998): 153.
5 More on the post-war Vichy officials, including Papon, and their complicity in the Holocaust see: Eric Epstein, "Fit to Be Tried: Maurice Papon

and the Vichy Syndrome. Defeat and Collaboration," *Journal of Genocide Research* 1, no. 1 (March 1999): 115.

6 Cole, "Massacres and Their Historians," 118.

7 Hannah Arendt, "Why the Crémieux Decree Was Abrogated," *Contemporary Jewish Record*, April 1, 1943, 119, 121.

8 Jeffrey C. Isaac, *Arendt, Camus, and Modern Rebellion* (New Haven, CT: Yale University Press, 1992); Ned Curthoys, "The Refractory Legacy of Algerian Decolonization: Revisiting Arendt on Violence," in *Hannah Arendt and the Uses of History. Imperialism, Nation, Race, and Genocide*, ed. Richard King and Dan Stone (New York, Oxford: Berghahn Books, 2007), 109–29.

9 Another explanation has it, as was explained in Chapter 2, that Arendt was outright Eurocentric, and perhaps even racist, hence her sensitivity to the deprivation and dispossession of the peoples of the south did not match her responsiveness to the supremacy of the political ideals of the north. In particular see Kathryn T. Gines's *Hannah Arendt and the Negro Question* (Bloomington: Indiana University Press, 2014).

10 David Macey, *Frantz Fanon: A Biography* (London; New York: Verso, 2012), 258. Macey quotes Jean Cohen.

11 Macey, 271.

12 Maurice Merleau-Ponty, *Signes* (Paris: Folio, 2001), 547.

13 Elizabeth Frazer and Kimberly Hutchings, "On Politics and Violence: Arendt Contra Fanon," *Contemporary Political Theory* 7, no. 1 (February 1, 2008): 90–108.

14 Annabel Herzog, "Hannah Arendt's Concept of Responsibility," *Studies in Social and Political Thought* 10 (August 2004): 39–56.

15 Herzog, 42.

16 Anthony C. Alessandrini, *Frantz Fanon and the Future of Cultural Politics: Finding Something Different* (Lanham, MD: Lexington Books, 2014), 163.

17 Frantz Fanon, *Toward the African Revolution. Politcal Essays*, trans. Haakon Chevalier (New York: Grove Press, 1964), 44. A different translation of the original text is also possible: "La fin du racisme commence avec cette soudaine incompréhension. La culture spasmée et rigide de l'occupant, libérée s'ouvre enfin à la culture du peuple devenue réellement frère. Les deux cultures peuvent s'affronter, s'enrichir. En conclusion, l'universalité réside dans cette décision de prise en charge du relativisme réciproque de cultures différentes une fois exclu irréversiblement le statut colonial." See Frantz Fanon, "Racisme et Culture," *Présence Africaine*, no. 8/10 (1956): 131.

18 Leo Zeilig, *Frantz Fanon: The Militant Philosopher of Third World Revolution* (London, New York: I.B. Tauris, 2016), 247.

19 Christopher J. Lee, *Frantz Fanon: Toward a Revolutionary Humanism* (Athens: Ohio University Press, 2015), 179–94.

20 Edward W. Said, *Reflections on Exile and Other Essays* (Cambridge, MA: Harvard University Press, 2000), 452.

21 Herzog, "Hannah Arendt's Concept of Responsibility," 42.

22 According to Bhabha, she is "at best, only half right" because Fanon is more cautious than she understood. Homi K. Bhabha, 'Framing Fanon', in *The Wretched of the Earth* (New York: Grove Press, 2007), XXXV.

23 Patricia Owens, *Between War and Politics: International Relations and the Thought of Hannah Arendt* (Oxford: Oxford University Press, 2007), 9.

24 Jurgen Habermas and Thomas McCarthy, "Hannah Arendt's Communications Concept of Power," *Social Research* 44, no. 1 (1977): 8.
25 Kathryn T. Gines, "Arendt's Violence/Power Distinction and Sartre's Violence/Counter-Violence Distinction: The Phenomenology of Violence in Colonial and Post-Colonial Context," in *Phenomenologies of Violence*, ed. Michael Staudigl (Leiden: Brill, 2013), 124–25.
26 Jens Hanssen, "Translating Revolution: Hannah Arendt in Arab Political Culture," *Journal for Political Thinking HannaArendt.net* 7, no. 1 (2013), accessed March 28, 2019, www.hannaharendt.net/index.php/han/article/view/301.
27 Barbara Deming, *Revolution & Equilibrium* (New York: Grossman Publishers, 1971), 197.
28 Fanon, 'Racisme et Culture', 131.
29 On the centrality of land to colonial expansion and elimination of the native population see Patrick Wolfe, "Settler Colonialism and the Elimination of the Native," *Journal of Genocide Research* 8, no. 4 (December 1, 2006): 387–409.
30 Ato Sekyi-Otu, *Fanon's Dialectic of Experience* (Cambridge, MA: Harvard University Press, 1997), 164.
31 Zeilig, *Frantz Fanon*, 226–27.
32 Tsenay Serequeberhan, *The Hermeneutics of African Philosophy: Horizon and Discourse* (New York: Routledge, 1994), 77.
33 Habermas and McCarthy, 'Hannah Arendt's Communications Concept of Power', 16.
34 Hannah Arendt, "The Jewish Army - the Beginning of a Jewish Politics?" in *The Portable Hannah Arendt*, ed. Peter Baehr (Harmondsworth, England: Penguin Books, 2000), 46–8.
35 The international form of national consciousness, proposed by Fanon, resembles what the European Union was capable of achieving: uniting national consciousnesses into a European one, without diminishing the allure of what nation states have to offer culturally, historically, and even ideologically. Such continent-wide consciousness requires, however, further thoughts on how it relates to other non-European consciousnesses.
36 Macey, *Frantz Fanon*, 257.
37 Haythem Guesmi, "Au Maghreb, le racisme contre les Noirs persiste," *Courrier international*, December 8, 2017, accessed March 28, 2019, www.courrierinternational.com/article/au-maghreb-le-racisme-contre-les-noirs-persiste.
38 One party systems have been factually in place and despite the nominally democratic systems of government in many Arab countries since independence. The National Democratic Party of Egypt (*Al-Hizb al-wataniyy ad-dimuqratiyy*) has ruled unilaterally in Egypt since late 1970s until 2011 despite the formal pluralist political scene. The situations was the same in Tunisia (*At-tagammu' ad-dusturiyy ad-dimuqratiyy*, Democratic Constitutional Rally ruled since 1956 until 2011), Syria (*Hizb al-ba'th al-'arabiyy al-ishtirakiyy*, Arab Socialist Baath Party) and many other countries.
39 Emmarie Deputy, "Designed to Deceive: President Hosni Mubarak's Toshka Project," Thesis (University of Texas, 2011), accessed March 28, 2019, https://repositories.lib.utexas.edu/handle/2152/ETD-UT-2011-05-3121.

40 Moutaz Ali, "The Eighth Wonder of the World?" *Qantara.de*, March 7, 2017, accessed March 28, 2019, https://en.qantara.de/content/libyas-great-man-made-river-irrigation-project-the-eighth-wonder-of-the-world.
41 "Egypt: Hundreds Disappeared and Tortured amid Wave of Brutal Repression," *Amnesty International*, accessed May 20, 2018, www.amnesty.org/en/latest/news/2016/07/egypt-hundreds-disappeared-and-tortured-amid-wave-of-brutal-repression/.
42 "Egypt Police Jailed over 2010 Death of Khaled Said," *BBC News*, March 3, 2014, accessed March 28, 2019, www.bbc.com/news/world-middle-east-26416964.
43 Following the pan-Arab initiatives by Gamal Abdel Nasser, including an Arab union with Syria in late 1950s, the common Arab project has proven politically incapable of presenting a unified front and failed at promoting the single joint Arab interest: independent Palestine. For analysis of internal Arab divisions see: Robert Springborg, "1967 to 2017: Arab Unity, a Fickle Beast," *alaraby*, accessed 20 May 2018, www.alaraby.co.uk/english/comment/2017/6/9/1967-to-2017-arab-unity-a-fickle-beast; "Arab Unity: The End?" *Al Jazeera*, accessed 20 May 2018, www.aljazeera.com/focus/arabunity/2008/03/200861517319872300.html.
44 Frantz Fanon, "Fondement Réciproque de La Culture Nationale et Des Luttes de Libération," *Présence Africaine*, no. 24/25 (1959): 88.
45 Quotation continues: "This unease was the reason Arendt grew disenchanted with Zionism even before the state of Israel was declared, and it was the reason why Said left the Palestinian National Council in 1991." See Jens Hanssen, "Translating Revolution: Hannah Arendt in Arab Political Culture," *Journal for Political Thinking HannaArendt.net* 7, no. 1 (2013), accessed March 28, 2019, www.hannaharendt.net/index.php/han/article/view/301.
46 Hannah Arendt, "Imperialism, Nationalism, Chauvinism," *The Review of Politics* 7, no. 4 (1945): 450.
47 Hannah Arendt, *Crises of the Republic: Lying in Politics, Civil Disobedience, On Violence, Thoughts on Politics and Revolution* (HMH, 1972), 49–102.

Bibliography

Alessandrini, Anthony C. *Frantz Fanon and the Future of Cultural Politics: Finding Something Different*. Lanham, MD: Lexington Books, 2014.

Ali, Moutaz. "The Eighth Wonder of the World?" *Qantara.de*, 7 March 2017. Accessed March 28, 2019. https://en.qantara.de/content/libyas-great-man-made-river-irrigation-project-the-eighth-wonder-of-the-world.

"Arab Unity: The End?" *Al Jazeera*. Accessed 20 May 2018. www.aljazeera.com/focus/arabunity/2008/03/200861517319872300.html.

Arendt, Hannah. *Between Past and Future*. New York: Penguin Classics, 2006.

———. *Crises of the Republic: Lying in Politics, Civil Disobedience, On Violence, Thoughts on Politics and Revolution*. San Diego; New York; London: HMH, 1972.

———. "Imperialism, Nationalism, Chauvinism." *The Review of Politics* 7, no. 4 (1945): 441–63.

———. *On Revolution*. London: Penguin Books, 1990.

———. *On Violence*. Orlando; Austin; New York; San Diego; London: HMH, 1970.
———. *The Human Condition*. Chicago, IL; London: The University of Chicago Press, 1998.
———. "The Jewish Army—the Beginning of a Jewish Politics?" In *The Portable Hannah Arendt*, edited by Peter Baehr, 46–8. Harmondsworth, England: Penguin Books, 2000.
———. *The Origins of Totalitarianism*. Orlando; Austin; New York; San Diego; London: HMH, 1973.
———. "We Refugees." In *Altogether Elsewhere. Writers on Exile*, edited by Marc Robinson, 110–9. Boston, London: Faber and Faber, 1996.
———. "Why the Crémieux Decree Was Abrogated." *Contemporary Jewish Record*, 1 April 1943, 115–23.
Bhabha, Homi K. "Framing Fanon." In *The Wretched of the Earth*, New York: Grove Press, 2007.
Cole, Joshua. "Massacres and Their Historians: Recent Histories of State Violence in France and Algeria in the Twentieth Century." *French Politics, Culture & Society* 28, no. 1 (Spring 2010): 106–26.
Curthoys, Ned. "The Refractory Legacy of Algerian Decolonization: Revisiting Arendt on Violence." In *Hannah Arendt and the Uses of History. Imperialism, Nation, Race, and Genocide*, edited by Richard King and Dan Stone, 109–29. New York, Oxford: Berghahn Books, 2007.
Deming, Barbara. *Revolution & Equilibrium*. New York: Grossman Publishers, 1971.
Deputy, Emmarie. "Designed to Deceive: President Hosni Mubarak's Toshka Project." Thesis, University of Texas, 2011. Accessed March 28, 2019. https://repositories.lib.utexas.edu/handle/2152/ETD-UT-2011-05-3121.
"Egypt: Hundreds Disappeared and Tortured amid Wave of Brutal Repression." *Amnesty International*. Accessed May 20, 2018. www.amnesty.org/en/latest/news/2016/07/egypt-hundreds-disappeared-and-tortured-amid-wave-of-brutal-repression/.
"Egypt Police Jailed over 2010 Death of Khaled Said." *BBC News*, March 3, 2014. Accessed March 28, 2019. www.bbc.com/news/world-middle-east-26416964.
Epstein, Eric. "Fit to Be Tried: Maurice Papon and the Vichy Syndrome. Defeat and Collaboration." *Journal of Genocide Research* 1, no. 1 (March 1999): 115.
Fanon, Frantz. "Fondement Réciproque de La Culture Nationale et Des Luttes de Libération." *Présence Africaine*, no. 24/25 (1959): 82–9.
———. "Racisme et Culture." *Présence Africaine*, no. 8/10 (1956): 122–31.
———. *The Wretched of the Earth*. New York: Grove Press, 2007.
———. *Toward the African Revolution. Political Essays*. Translated by Haakon Chevalier. New York: Grove Press, 1964.
Frazer, Elizabeth, and Kimberly Hutchings. "On Politics and Violence: Arendt Contra Fanon." *Contemporary Political Theory* 7, no. 1 (1 February 2008): 90–108.

Gines, Kathryn T. "Arendt's Violence/Power Distinction and Sartre's Violence/Counter-Violence Distinction: The Phenomenology of Violence in Colonial and Post-Colonial Context." In *Phenomenologies of Violence*, edited by Michael Staudigl, 123–44. Leiden: Brill, 2013.

Golsan, Richard J. "Memory's Bombes a Retardement: Maurice Papon, Crimes against Humanity, and 17 October 1961." *Journal of European Studies* 28, no. 1/2 (March 1998): 153.

Guesmi, Haythem. "Au Maghreb, le racisme contre les Noirs persiste." *Courrier International*, 8 December 2017. Accessed March 28, 2019. www.courrierinternational.com/article/au-maghreb-le-racisme-contre-les-noirs-persiste.

Habermas, Jurgen, and Thomas McCarthy. "Hannah Arendt's Communications Concept of Power." *Social Research* 44, no. 1 (1977): 3–24.

Hanssen, Jens. "Translating Revolution: Hannah Arendt in Arab Political Culture." *Journal for Political Thinking HannaArendt.net* 7, no. 1 (2013). Accessed March 28, 2019. www.hannaharendt.net/index.php/han/article/view/301.

Herzog, Annabel. "Hannah Arendt's Concept of Responsibility." *Studies in Social and Political Thought* 10 (August 2004): 39–56.

Isaac, Jeffrey C. *Arendt, Camus, and Modern Rebellion*. New Haven, CT: Yale University Press, 1992.

Lee, Christopher J. *Frantz Fanon: Toward a Revolutionary Humanism*. Athens: Ohio University Press, 2015.

Lindqvist, Sven. *'Exterminate All the Brutes': One Man's Odyssey into the Heart of Darkness and the Origins of European Genocide*. Translated by Joan Tate. New York: The New Press, 1997.

Macey, David. *Frantz Fanon: A Biography*. London; New York: Verso, 2012.

Merleau-Ponty, Maurice. *Signes*. Paris: Folio, 2001.

Owens, Patricia. *Between War and Politics: International Relations and the Thought of Hannah Arendt*. Oxford: Oxford University Press, 2007.

Planche, Jean-Louis. *Sétif 1945. Chronique d'un massacre annoncé*. Paris: Perrin, 2010.

Said, Edward W. *Reflections on Exile and Other Essays*. Cambridge, MA: Harvard University Press, 2000.

Sekyi-Otu, Ato. *Fanon's Dialectic of Experience*. Cambridge, MA: Harvard University Press, 1997.

Serequeberhan, Tsenay. *The Hermeneutics of African Philosophy: Horizon and Discourse*. New York: Routledge, 1994.

Springborg, Robert. '1967 to 2017: Arab Unity, a Fickle Beast'. *alaraby*. Accessed 20 May 2018. www.alaraby.co.uk/english/comment/2017/6/9/1967-to-2017-arab-unity-a-fickle-beast.

Wolfe, Patrick. "Settler Colonialism and the Elimination of the Native." *Journal of Genocide Research* 8, no. 4 (1 December 2006): 387–409.

Zeilig, Leo. *Frantz Fanon: The Militant Philosopher of Third World Revolution*. London, New York: I.B. Tauris, 2016.

4 Jihadism and other remains of colonisation

> Fatalism relieves the oppressor of all responsibility since the cause of wrong-doing, poverty, and the inevitable can be attributed to God.
>
> Frantz Fanon (WotE, 17)

Jihadism – as Jarret M. Brachman suggests – is a "clumsy and controversial term"[1] but it is widely used in the public discourse. Here, it means an extremist ideology that calls for violence against all those who do not abide by a terror group laws, which the group claims are the only proper interpretation of Islam. In such a form jihadism is the tyranny that Arendt warned of appearing when the public space shrinks and bureaucratisation spreads. It is also the fundamentalism and "Dark Ages" that grow in the Middle East when self-creation as political subjects fails, of which Fanon warned. Jihadism is a contemporary violent ideology that interestingly fits the conditions both in the south and the north. With the decrease in leftist, communist and liberation terror, radicalisation – which is a natural social phenomenon – is becoming "Islamised" in the words of Roy.[2] Jihadism has become the prevalent form of non-state political violence in Europe although separationist and far-right terror is still high. Europol report on terrorism in 2018 shows that even though ethno-nationalist and separatist terrorist attacks continue to far outnumber attacks inspired by any other ideology, jihadist attacks have been the most lethal and impactful on the societies.[3] In this chapter a matrix for the reasons of jihadist appeal in the Middle East (south) and Europe (north) will be given in relation to the factors that push to violence as enumerated by Fanon and Arendt.

Radical comes from "root" (*radix* in Latin) and could mean going back to the roots or a change starting from the roots. In the political context major dictionaries focus on the latter defining a radical as someone who desires extreme change of part or all of the social order.[4] Groups or movements that have such an extreme change for their goal abound[5] but none is as international, potent, conspicuous and violent today as the jihadists. The jihadists cannot be called radicals – they are violent extremists who use religious slogans as legitimisation for their violence but their violence is essentially political, pertaining to the public sphere, and should be seen as such. As Haupt and Weinhauer note about religiously-inspired violence "it is hard to discern religious motives from political ones, and religion does not cause terrorism but serves to legitimise it."[6]

The plethora of reasons for jihadi groups' local and global appeal, the lack of any one single set of political, cultural or economic push factors, the diversity of backgrounds of supporters of radicalism, the multitude of languages and skin colours of foreign fighters, corroborate the thesis that violent Islamic extremism or jihadism has become already or has the potential of becoming a general, global, radical anti-systemic movement,[7] with the rejection of the outside world as its basic ideological dogma. Among the radical groups that draw young Arabs and Europeans alike ISIS (the "Islamic State of Iraq and Syria" or the so called "Islamic State") stands out as the most successful radical group globally. Even if ISIS's state-forming project is now largely defeated, the group together with Al-Qaeda had been very successful in attracting followers. There are two major reasons for this: it created a physical entity (the so-called Caliphate) and a global imagined community by emphasising that its call is addressed to every Muslim and – as will be shown – to anyone prone to such a call:

> O Muslims everywhere (…). Raise your head high, for today – by Allah's grace – you have a state and khilafah, which will return your dignity, might, rights, and leadership. It is a state where the Arab and non-Arab, the white man and black man, the easterner and westerner are all brothers. It is a khilāfah that gathered the Caucasian, Indian, Chinese, Shami, Iraqi, Yemeni, Egyptian, Maghribi (North African), American, French, German, and Australian. Allah brought their hearts together, and thus, they became brothers by His grace, (…) defending and guarding each other (…) Their blood mixed and became one.[8]

In that ISIS is specifically different from its predecessor – Al-Qaeda – which focused on fighting the enemies without building a community or institutions. ISIS, however, directly stems from Al-Qaeda in Iraq and both share similarities in their stages of development. Al-Qaeda had once been confined to the Af-Pak region but in 2006 and 2007 entered its mushrooming stage attracting affiliates outside of its core territory of conduct (i.e. Al-Qaeda in the Islamic Maghreb or Al-Shabaab in Somalia). Likewise, having established its base in Syria and Iraq in 2014, ISIS began to attract many local jihadi groups in the Muslim world outside of the core territory (i.e. Boko Haram in Nigeria, Wilayat Barqa in Libya, Jund al-Khilafa in Algeria or Wilayat Sinai in Egypt). They plead allegiance to ISIS, because they too want to wield sovereignty over their lands against local enemies. ISIS also serves as a well-known brand that enjoys greater global attention today than Al-Qaeda and the reasons for its attractiveness go beyond the lure any other jihadi organisation, making it an apt case to study the reasons for extremist groups' appeal comprehensively.

ISIS and other jihadists fill the vacuum left by other radical revolutionary movements, which gradually lost their popularity after the end of the cold war. There have always been individuals alienated in the systems they lived in, and their radicalisation has always been a result of an amalgam of reasons: personal, sociological, political and economic. Looking at ISIS today as a phenomenon limited to one religion or background misses its gradual but advancing transformation into a global anti-systemic movement that can attract anyone in the south and north:[9] a member of a family of a former jihadist in Libya, a British teenage woman, a Chechen opposition activist from Grozny, a former Baath party official in Iraq, a WASP student from California or a French convert from the countryside. Reports show that jihadism has attracted more and more people, particularly more women, who are young and active online – only one in ten Western fighters joined a group other than ISIS.[10]

How big is jihadist appeal?

There is no more popular violent extremist group than the ISIS/Al-Qaeda couple. At peak times (December 2015) there were some 30,000 foreign fighters in Syria and Iraq who came from 86 countries. Most of them hailed from the Middle East and the Maghreb (more than 16,000). An estimated 5,000 nationals from Western Europe had joined the ranks of extremists in Syria and Iraq, which marked a 100% increase from 2,500 in June 2014, while 4,700 come from former Soviet republics – an increase by more than 300%.[11] Top nationalities among the fighters included: Tunisians (6,000), Saudis (2,500), Russians (2,400), Turks (2,100), Jordanians (2,000). The majority of

European fighters come from four countries: France (1,700), United Kingdom (760), Belgium (470) and Germany (470).[12] Any quantification of ISIS' or other Islamist radical groups' appeal, meaning any kind of support for or attraction to ISIS, not necessarily expressed by taking up arms and going to Syria, cannot be precise. Only an approximated conclusion can be drawn from the available data.

In relative terms and at the height of its territorial project in Syria Islamic State may have enjoyed greater popularity in Europe than in the Middle East: the highest number of sympathisers could be found in France (about 16%), the UK (about 7%) and in Germany (3%).[13] In France the support was rising with age: 4% of 18- to 24-year-olds sympathised with ISIS, compared to 6% of 24- to 35-year-olds and 11% of 35- to 44-year-olds. Just 3% of Egyptians expressed a positive opinion of the organisation, 5% of Saudis and under 1% of Lebanese.[14] A "VOICES" study (run by Italian academics) of 2 mln posts showed that support for ISIS was stronger in Arabic social media in Europe than in Syria.

Outside Syria, support for ISIS, always a minority among online communities, rose significantly. Forty-seven per cent of studied tweets and posts from Qatar, 35% from Pakistan, 31% from Belgium and almost 24% of posts from UK and 21% from the US were classified as being supportive of the jihadist organisation compared with just under 20% in Jordan, Saudi Arabia (19.7%) and Iraq (19.8%).[15]

However, AlJazeera Arabic ran an online poll a year after ISIS captured Mosul in 2014, which showed that 81% (more than 46,000) of the voters supported the organisation's advances in Syria and Iraq versus 19%.[16] If there can be any concrete conclusion drawn from these polls it is that ISIS does have an unusually large for a terrorist organisation following, and it is not limited to the Middle East. Jihadism attracts followers both in the south and north.

There is rich literature about radicalisation and its reasons,[17] which in the case of ISIS popularity in Europe and the Middle East can be grouped in four categories that encompass Fanon's and Arendt's reasons for the appeal of violence: (1) ideological and political, (2) sociological and psychological, (3) economic and (4) practical. In each of the categories a distinction between Arab (southern) and European (northern) citizenry needs to be made as the two national groups are often drawn to ISIS by different factors. Overall, the reasons, even though categorised and ordered, are not disconnected from one another, they merge into one push and pull factor, sometimes conditioning one another.

Plastic Islam and utopia: ideological and political reasons behind the attractiveness of jihadism

The relation of a violent extremist ideology to a religion can be compared to the relation of criminality to a society, or human trafficking to globalisation. Jihadism is possible in Islam – like other violent extremisms are possible in other religions – but its agenda is not religious, it is social and political. What is there in Islam that jihadism is able to exploit?

There is no one Islam. As a religion Islam is simple: following the five pillars is enough to be a good Muslim. Without a complex and dogmatic process it makes excommunication virtually impossible. Islam is also ideologically flexible, more than other monotheistic religions. One of the most acclaimed sociologists and political scientists, Ernest Gellner, claimed that Islam is "the most protestant of the great monotheisms, it is ever Reformation-prone (Islam could indeed be described as Permanent Reformation)."[18] It has undergone many successive self-reformations – Gellner calls it Islamic Protestantism[19] – the urge to reform has always been present in Islam.[20] Jihadism and ISIS are therefore not unexpected – historically they were preceded by other revolutionary intra-religious movements. The ability to revolt within comes, among other reasons, from the dogmatic absence of clergy in Sunni Islam (90% of all Muslims), meaning any intermediary between man and God, and from the multitude of possible interpretations of the scripture. The egalitarian, direct relations to God is attractive as a concept but it also allows self-proclaimed, uncontrollable imams or those from outside of the mainstream official Islam to be seen by the believers as religious authority. It weakens any religious oversight and facilitates revolt and recruitment. So does the easiness of being a good Muslim: if you follow the five pillars of Islam you will go to heaven, without having to confess to anyone. The universal features of Islam – simplicity, combined with the freedom to judge for oneself and the "ambiguity of concrete moral and political precepts"[21] – make for a religion one can mould into almost anything that fits their needs. Jihad, too, even if the word itself is Arabic, has a universal meaning as a revolution, fight against the oppressor or internal, personal effort to be good.

Jihad per se is no longer restricted to Muslims. Factually becoming Muslim is of secondary importance, it is a formal and last step to be included in the community – a symbolic ceremony present in many brotherhood-like systems: like a scout oath, or swearing allegiance to any leader or law in the army. In France jihadi recruits for 20 years have hailed either from second generation of migrants or organically ("de souche") French converts.[22] American recruits, many of them converts, are Caucasian, Somali-American, Vietnamese-American, Bosnian-American, Arab-American, among other ethnicities and nationalities."[23]

The anti-systemic nature of ISIS is corroborated in its opposition to the international system and to the traditional Westphalian state system in the Middle East, an order that is thought un-Muslim, created and oriented by the West and the colonist. The Middle Eastern state has used violence indiscriminately, being particularly oppressive toward Islamists and impotent in providing services to the people in general. ISIS used to be the strongest in those provinces in Syria and Iraq where the central government had been the weakest. Several scholars have pointed to mutually reinforcing relation between authoritarianisms (state violence/oppression) and extremisms (non-state violence).[24]

Vast majority of resistance movements since 1970s in the Middle East have been religious in nature. Jihadists can be against state institutions but in their ideology Islam plays a similar role to that of nationalisms in the West – in it the caliphate is presented as an alternative, moral, responsible and own state. Like in the Middle East, in the West personal grievances can also be easily directed against the state or system. The more ostentatious it is in its secularity, the greater sense it makes to become religious in opposition to it. In France, young Muslims, disconnected from the culture of the Middle East, often without knowledge of Arabic,[25] can shape their beliefs practically at will, thanks to malleability of Islam.[26] Foreign and local fighters within the ranks of jihadist groups very often have poor knowledge of Islam[27] – the medieval Quranic text is difficult to comprehend and interpret even for the educated.

For Muslims ideological strength of ISIS also comes from the fact that jihadists play on the sentimental and communal analogies between today and the time of early Islam in the Middle Ages. Back in the 7th century it was created among Christians, Jews and other denominations, it found its way – the "right one" – also by converting those around to the "righteous," final religion (Muhammad is the "seal" of prophecies). Jihadism calls Muslims, like Mohammad did in 7th century, to perform a hijra (emigration) from the land of the disbelievers to the land of Islam.[28] Historical narrative is important, as Fanon noted, because it epitomised the longing for a great victorious past, when civilisational achievements of the Middle East exceeded those of Europe.

On top of that, recent policy failures have also contributed to today's appeal of jihadism. The past four decades form a series of mistakes in foreign policy by the United States, and recently – Europe as well. Fanon's remark that "The reserves of colonialism are far richer and more substantial than those of the colonized" (WoTE, 90) got its validation. In the 1980s the United States helped jihadists in Afghanistan and pitted the regime of Saddam Hussain against the Iranians, in 1990s they intervened twice in Kuwait and Iraq (though, wisely, stopped short of deposing Hussain), in 2000s they invaded Afghanistan and Iraq again, destroying the social and political fabric of both, in 2011 together with UK and France they helped to decompose Libya. Those mistakes empowered both radicals and authoritarian governments, and they also remain a potential radicalisation factor for the future. Political consciousness of the 15- to 30-year-olds in the Middle East and, to large extent in Europe as well, witnessed mostly these events from the history of the world, hence, is easily shaped by anti-Americanism and anti-Westernism in general.[29]

Additionally, jihadists' allure also comes from the presumed higher moral ground, given by religious piety, that counters the "immorality" of the West.[30] Their moral system was combined with a state-like structure (caliphate ministries, passports, money), culture and art (papers, movies and an Orwellian-like newspeak) to form a complete utopia. In this sense "religion matters because it is a legitimizing resource of real potency, and, in the hands of innovating ideologists, provides moral justifications for violence."[31]

In such a system traditional political affiliations (left or right) were either irrelevant or secondary. Muslim fundamentalists should be in European political terms considered far-right wing with their cultivation of tradition, submission to one religion, social rigorism and disdain of minorities. But they attract right-wingers[32] and left-wingers with its anti-imperialist rhetoric alike.[33]

Cross-cultural anti-systemism and exclusion: sociological and psychological reasons for the appeal of jihadism

There are striking ideological and background similarities between ISIS today and radical leftist utopias that emerged among the first post-Second World War generation in Western Europe. Both movements had a revolution for their goal: the leftist radicals wanted to end labour alienation, while the Islamic radicals want to end Muslim alienation – both hated their own governments and the US. Like ISIS today, the Baader-Meinhoff group was also a symptom of a crisis of ideologies and clashing values: back then it was socialism/communism and capitalism that divided Europe in two, with Germany at the heart of the division. Today, more in Western Europe than in the Middle East, ISIS breeds on the paradoxes of ultra-liberal social order that allows almost any form of individualisation but fails to accommodate religiosity, incite community and expand the public sphere in general. In the south a clash of values is similar and at least as visible with a dominant role of the corrupt elite, either showing disdain for Islam and the poor or a nominal and rhetorical embracement of religion in order to use it for political purposes. Such hypocrisy clashes with the religiosity of the people.

Analogies between European utopian radicals and jihadism today extend to social conditions and psychological disposition. After the 1968 violent events that Arendt analysed, violent extremism in 1970s raised its head when the first economic and values crisis struck Europe after the golden age of 50s and 60s.

> The urge to bring the architecture of security and stability crashing down on the heads of their parents' generation was the extreme expression of a more widespread scepticism, in the light of the recent past, about the local credibility of pluralist democracy.[34]

Today again, after the golden age of the 90s an economic and perhaps a social crisis bite Europe again. It consists of a cultural crisis that dissolves the values of Europe and a social backlash against the liberalism of the parents' generation. That crisis today also debilitates non-religious peoples' understanding of religion and its "ultimate concern"[35] function – how a set of beliefs becomes an internal imperative above worldly life.

The modern, ultraliberal systems and societies may have cut the branch they had been built on – religion, community, public space etc. – ridding individuals of a foundation that not only the shaky souls need.[36] This drive of many Europeans towards ISIS resembles what Erich Fromm famously called "escape from freedom." According to Fromm, freedom is so alienating and demanding that people would rather escape into authoritarian, totalitarian systems just to feel safe again. When Arendtian hypocrisy makes it impossible to identify the responsible party, the system feels closed and undemocratic, its institutions – dead. Undoubtedly, crisis of personality and frustration are primary psychological reasons that push Europeans and Middle Easteners alike into ISIS arms.[37] It is not uncommon for second immigrant generation to have identity disorders[38] and feel rejected by societies they live in. The caliphate satisfies the need to belong, mean and feel safe again, albeit pathologically.

Another social and psychological reason that most probably binds the two sides of the Mediterranean in making ISIS attractive is that it revolts against the older generation. The generational gap always makes youth rebel but in recent decades the gap may have become deeper and broader primarily owing to the change in communication means. Most foreign fighters from Europe are in their 20s,[39] while in general most of the Arab world is very young. In its global aspect it is a revolution of the young. The generational gap came to the limelight in the revolts of 2011 already. With the crackdown both on religious and civil activism in many Arab countries today and a deep generational gap in place, there are simply more young people (in absolute numbers and proportionally) today that are prone to ISIS propaganda than in the past.

Relative deprivation and economic reasons for the popularity of jihadism

The caliphate attracts people who are economically deprived but not exclusively. To understand why and before absolute economic deprivation can be analysed, it is important to see the reasons behind the appeal of radicals such as ISIS that can be located on the overpass between society and economy, namely relative deprivation (RD). The concept is defined as a person's perceived disparity between their capabilities and expectations.[40] In other words one expects more than one achieves, regardless of the objective status of the person in question. The expectations today, built up by globalised mass media, greater inequalities, a perception that so many people have so much more and their lives are presumably so much better, make societies prone to political disruptions or even violence.

> Deprivation is relevant to the disposition to collective violence to the extent that many people feel discontented about the same things. Unexpected personal deprivations such as a failure to obtain an expected promotion or the infidelity of a spouse ordinarily affect few people at any given time and are therefore narrow in scope. Events and patterns of conditions like the suppression of a political party, a drastic inflation, or the decline of a group's status relative to its reference group are likely to precipitate feelings of RD among whole groups or categories of people and are wide in scope.[41]

With regard to societies today, it seems that both personal RD and whole group RD can be causes of ISIS appeal among individuals who experienced unexpected personal deprivation (job loss, heartbreak, etc.) and those who feel it collectively (glass ceiling for migrant descendants in France, crackdown on Islamist parties in the Middle East, etc.) thanks to the specific communal characteristic of ISIS. Violence committed outside of the land of this imagined community (ISIS territory in Syria and Iraq) acquires features of collective violence.

When looking at causes of ISIS appeal the category of absolute economic deprivation may still be valid (although in a different sense) in some case with reference to the conquered population in Syria and Iraq, as well as recruits from the Middle East and the Caucasus. In Europe this reason may comparatively play a lesser role, although undoubtedly European recruits come from lower social strata, often with criminal record. In agreement with Fanon's assessment from the 1960s more than half of a century, Arab states are petrified class societies with a long history of Westernised elite rule, meagre social mobility and a large portion of Arab societies living under the national poverty levels. The UNDP Arab Human Development Report acutely enumerated the economic shortcomings of the Arab world in subsequent reports, finding i.a. that the percentage of populating living under the national poverty line reached 30% in Syria and Lebanon, 41% in Egypt, 59.5% in Yemen, with a total of around 40% for the whole region in 2009, which equalled 65 million people.[42] The same report alarmed about the rising inequalities, insecurity of disadvantaged individuals and joblessness with a potential to grow into a genuine regional plague if 51 million jobs were not created by 2020.[43] The same applies to youth in Europe, where inequalities and unemployment frustrate as many as half of all young population of Spain for example.[44] Additionally, the past couple of years created wars in Libya, Yemen, Syria and Iraq, which put many more people at risk.

The Middle Eastern region is particularly prone to climate change that results in droughts, water shortage, desertification and urban pressure as masses move from rural to urban areas. ISIS capital – Raqqa – was one of the worst hit by drought provinces of Syria between 2007 and 2011. The drought caused water shortages, agricultural failures, livestock mortality and, hence, pushed 1.5 million people to migrate to the peripheries of urban centres.[45] In Europe these reasons may cause more irregular migration from the Middle East and Africa and greater hostility towards the incomers.

Practical and technical reasons for jihadist outreach

The modern technologised way of living also makes jihadism appealing. ISIS benefited immensely from the coincidence of young people using new technological tools to communicate and entertain themselves. The young are the most skilful group to be online, getting the most from traditional operating systems, social media, smartphone apps and dark web. Groups like ISIS are also prolific online, with thousands of Twitter and Facebook accounts at peak times that constantly targeted disaffected youth in the West. They use apps and other programs in recruitment and dissemination of propaganda.

Hailing from more than 80 countries they have multi-linguistic and wide-outreach dissemination resources at their disposal. Many statements used to be translated into more than five languages and disseminated on the most popular internet platforms, multiplying factual impact of ISIS. Its multi-lingual and hyper-modern production of clips and magazines "employs the aesthetic of contemporary Hollywood films, video games and TV shows"[46] built on action and brutality. The Hollywood-like character of these clips may either imply that terrorists want to target people in the West in particular or that this cultural style would simply be most effective since global mass culture is dominated by it anyway. By mimicking the so called massive multiplayer role playing games, like Call of Duty, they also target the young who play these games most frequently.

But this outreach would be much less effective if it wasn't for a personal bond that a recruiter develops with a potential recruit. Experts at the counter-radicalisation department (UCLAT) at the French Ministry of Interior have concluded that the true primary tool in recruitment is a personal relationship formed online. Such online bonds, in turn, naturally develop between younger people who spend more time online than in real social interactions. They are susceptible to taking online relations as more genuine or valuable than real life bonds. Arendtian fear of technology and Fanonian criticism of alienation reverberate in the technical and practical reasons for the attractiveness of jihadism.

* * *

Jihadism may be exclusively associated with the Muslim culture as a term and concept but it also has a universal, cross-cultural dimension and has become the ideology of some of the wretched of the earth today. This radical utopia serves as a global magnet for anti-systemism rather than just attracting religious locals in the Middle East. The selection of universal reasons for ISIS appeal and the level to which jihadism responds to the problems of southern and northern socio-political systems (Table 4.1) demonstrates that there is no single reason, nor a type of reasons that explains the popularity of ISIS in its entirety. Many of the reasons for its appeal will remain for years (economic problems, authoritarianism in the Middle East) and some, such as hypocrisy or the difficulty of reconciling different values crisis of values may grow. Perhaps the key to seeing an upward trend in the attractiveness of jihad as a global utopia is understanding that not only southern socio-political systems have deficiencies but, as Arendt teaches, Western culture is in political crisis too. The proportion of new foreign recruits from the Middle East to those from Europe may tilt in statistical favour of the latter ones, even if they will not be regular jihadi fighters but just people deprived of the possibility to be a political being. This greater radicalisation potential is present on both political extremes: the right and the left, as well as among disenfranchised social groups in Europe. The terrorist attacks and the rise of the right-wing parties make anti-immigrant rhetoric more prevalent. In turn, xenophobic narrative leads to more radicalisation.

Table 4.1 Features of contemporary societies and political systems, as identified by Arendt and Fanon, and the jihadist solutions to them

Problem	*Jihadist solution*
Unpredictability (HC)	Predictability
Rapid changes (HC)	Reversing changes
Hypocrisy (OV, WotE)	Truthfulness
Powerlessness (OV, WotE)	Powerfulness
Atomisation (OoT)	Community
Racism (OoT, WotE)	All-race approach
Injustice and inequality (OoT, WotE)	Justice and equality
Northern dominance (WotE)	Southern anti-systemism
Bureaucratisation – irresponsible and corrupt state apparatus (OV, WotE)	Responsibility
Deficiency of values and heritage, cultural void (WotE)	Protection of values and heritage
Rationalism (WotE)	Revolution and emotion

Notes

1 Jarret M. Brachman, *Global Jihadism: Theory and Practice* (London; New York: Routledge, 2008).

2 Olivier Roy, "Who Are the New Jihadis?" *The Guardian*, April 13, 2017, accessed 28 March 2019,_www.theguardian.com/news/2017/apr/13/who-are-the-new-jihadis.

3 "European Union Terrorism Situation and Trend Report 2018 (TESAT 2018)," *Europol*, accessed 28 March 2019, www.europol.europa.eu/activities-services/main-reports/european-union-terrorism-situation-and-trend-report-2018-tesat-2018.

4 Or "believing or expressing the belief that there should be great or extreme social or political change." See Cambridge Dictionary (http://dictionary.cambridge.org/dictionary/english/radical) and Encyclopedia Britannica (www.britannica.com/topic/radical-ideologist).

5 Radical groups differ: they can be violent or peaceful, have a leftist or a far-right ideology, or position themselves in the extreme to any other axis (i.e. radical ecologists).

6 Heinz-Gerhard Haupt and Klaus Weinhauer, "Terrorism and the State," in *Political Violence in Twentieth Century Europe*, ed. Donald Bloxham and Robert Gerwarth (Cambridge: Cambridge University Press, 2011), 205.

7 Olivier Roy discusses this thesis in his recent papers. See Olivier Roy, *Globalized Islam. The Search for a New Ummah* (New York: Columbia University Press, 2006); Olivier Roy, "Le djihadisme est une révolte générationnelle et nihiliste," *Le Monde*, January 8, 2016, accessed 28 March 2019, www.lemonde.fr/idees/article/2015/11/24/le-djihadisme-une-revolte-generationnelle-et-nihiliste_4815992_3232.html, Oliver Roy, Haoues Seniguer, "Comment l'islam est devenu la nouvelle idéologie des damnés de la planète," *Atlantico*, 4 July 2015, accessed 28 March 2019, www.atlantico.fr/decryptage/comment-islam-est-devenu-nouvelle-ideologie-damnes-planete-olivier-roy-haoues-seniguer-2221200.html.

8 "Abu Bakr al-Baghdadi's call," *Dabiq*, issue 1, 7, accessed July 12, 2017, http://media.clarionproject.org/files/09–2014/isis-isil-islamic-state-magazine-Issue-1-the-return-of-khilafah.pdf.

9 "This case [of an ISIS supporter], like others in communities across the United States and around the world, is an example of how a young person from any place and any background might make the terrible decision to try and become part of a terrorist organization" said a U.S. attorney. See "San Joaquin County Man Pleads Guilty to Attempting to Provide Material Support to ISIL," Press Release, Federal Bureau of Investigation, December 1, 2015, accessed March 28, 2019, www.fbi.gov/contact-us/field-offices/sacramento/news/press-releases/san-joaquin-county-man-pleads-guilty-to-attempting-to-provide-material-support-to-isil.

10 Courtney Schuster, David Sterman, and Peter Bergen, "ISIS in the West. The New Faces of Extremism," *New America Foundation*, November 15, 2015, accessed 28 March 2019, www.newamerica.org/international-security/future-war/policy-papers/isis-in-the-west/.

11 *Foreign Fighters: An Updated Assessment of the Flow of Foreign Fighters to Syria and Iraq* (The Saufan Group, December 2015), accessed March 28,

2019, http://soufangroup.com/wp-content/uploads/2015/12/TSG_Foreign-FightersUpdate_FINAL.pdf.

12 Ibid. Other nationalities notably represented: Austria (300), Bosnia (330), China (300), Egypt (1000), Indonesia (700), Kazakhstan (300), Kosovo (230), Lebanon (900), Morocco (1200), Netherlands (220), Sweden (300), Tajikistan (386).

13 Madeline Grant, "16% of French Citizens Support ISIS, Poll Finds," *Newsweek*, August 26, 2014, accessed March 28, 2019, www.newsweek.com/16-french-citizens-support-isis-poll-finds-266795; Jessica Elgot, "Islamic State Actually Has More Support In Britain Than In Arab Nations," *HuffPost UK*, October 16, 2014, accessed March 28, 2019, www.huffingtonpost.co.uk/2014/10/16/islamic-state-arab-nations-britain-support_n_5995548.html.

14 David Pollock, "ISIS Has Almost No Popular Support in Egypt, Saudi Arabia, or Lebanon," *WINEP*, October 14, 2014, accessed March 28, 2019, www.washingtoninstitute.org/policy-analysis/view/isis-has-almost-no-popular-support-in-egypt-saudi-arabia-or-lebanon.

15 Shiv Malik, "Support for Isis Stronger in Arabic Social Media in Europe than in Syria," *The Guardian*, November 28, 2014, accessed March 28, 2019, www.theguardian.com/world/2014/nov/28/support-isis-stronger-arabic-social-media-europe-us-than-syria.

16 "التصويت : هل تعتبر تقدم تنظيم الدولة الإسلامية في العراق وسوريا لصالح المنطقة؟" [At-Taswit: Hal ta'tabir taqaddum tanzim ad-dawla al-islamiyya fil-Iraq wa-Suriya li-salih al-mintaqa? Survey: Do you consider the advancement of the organization Islamic State to be in the interest of the region?], *Al Jazeera*, accessed June 3, 2018, www.aljazeera.net/votes/pages?voteid=5270.

17 For publications specifically interesting in the context of Islamist radicalization see i.e.: Farhad Khosrokhavar, *Inside Jihadism: Understanding Jihadi Movements Worldwide* (Abingdon, Oxon: Routledge, 2015); Thomas Olesen and Farhad Khosrokhavar, *Islamism as Social Movement* (Centre for Studies in Islamism and Radicalisation (CIR), Aarhus University, 2009), accessed March 28, 2019, www.ps.au.dk/fileadmin/site_files/filer_statskundskab/subsites/cir/pdf-filer/Hæfte2final.pdf; Tinka Veldhuis and Jørgen Staun, *Islamist Radicalisation: A Root Cause Model* (Netherlands Institute of International Relations Clingendael, 2009); Marina Ottaway, *ISIS – Many Faces, Different Battles* (Wilson Center, 2015), accessed March 28, 2019, www.wilsoncenter.org/sites/default/files/isis_many_faces_different_battles.pdf; John M. Venhaus, "Why Youth Join Al-Qaeda," United States Institute of Peace, May 2010, accessed March 28, 2019 www.usip.org/publications/2010/05/why-youth-join-al-qaeda.

18 Ernest Gellner, *Nations and Nationalisms* (Ithaca, NY: Cornell University Press, 2006), 77.

19 Gellner, 40.

20 Ernest Gellner, *Postmodernism, Reason and Religion* (London; New York: Routledge, 1992), 19.

21 Gellner, *Nations and Nationalisms*, 69.

22 Olivier Roy, "Le djihadisme est une révolte générationnelle et nihiliste," *Le Monde.fr*, November 24, 2015, accessed 28 March 2019, www.lemonde.fr/idees/article/2015/11/24/le-djihadisme-une-revolte-generationnelle-et-nihiliste_4815992_3232.html.

23 Schuster, Sterman, and Bergen, "ISIS in the West. The New Faces of Extremism."

24 Ronald Wintrobe, "Extremism, Suicide Terror, and Authoritarianism," *Public Choice* 128, no. 1/2 (2006): 169–95;_Marc Hetherington and Elizabeth Suhay, "Authoritarianism, Threat, and Americans' Support for the War on Terror," *American Journal of Political Science* 55, no. 3 (July 2011): 546–60; Asiem El Difraoui, "Authoritarianism and Radicalization towards Violent Extremism," in *Euromed Survey. Violent Extremism in the Euro-Mediterranean Region*, ed. Josep Ferré, 8th edition (Barcelona: IEMed, 2017), 34–39.

25 For an account of young Muslim civic leader's remarks about the older generation's Muslim leaders see American diplomatic cable 06PARIS6995 (https://wikileaks.org/plusd/cables/06PARIS6995_a.html, accessed March 28, 2019).

26 For example British Muslim radicals have a very basic or distorted understanding of Islam. See: Dina Temple-Raston, "New Terrorism Adviser Takes A 'Broad Tent' Approach," *NPR.org*, January 24, 2011, accessed March 28, 2019, www.npr.org/2011/01/24/133125267/new-terrorism-adviser-takes-a-broad-tent-approach.

27 "Journey of Young Africans into Violent Extremism Marked by Poverty and Deprivation," *UNDP*, accessed March 28, 2019, www.undp.org/content/undp/en/home/presscenter/pressreleases/2017/09/07/vers-l-extremisme-violent-en-afrique.html; "European Union Terrorism Situation and Trend Report 2018 (TESAT 2018)"; Zaman Al-Wasl, "Leaked ISIS Documents Reveal Most Recruits Know Little on Islam," *Haaretz*, August 16, 2016, accessed March 28, 2019, www.haaretz.com/middle-east-news/leaked-isis-documents-reveal-recruits-ignorant-on-islam-1.5424990.

28 "Lift your heads up high. You now have a state and a caliphate that restores your honor, your might, your rights and your sovereignty. The state forms a tie of brotherhood between Arab and non-Arab, white and black, Easterner and Westerner. The caliphate brings together the Caucasian, Indian, Chinese, Shami, Iraqi, Yemeni, Egyptian, north African, American, French, German and Australian… They are all in the same trench, defending each other, protecting each other and sacrificing for one another. Their blood mingles together under one flag [with] one goal and in one camp… perform hijra from darul-kufr to darul-Islam. There are homes here for you and your families." *Dabiq*, issue 1, 7.

29 Anti-Americanism also characterized the leftist radicals of the 70s in Germany and Italy, and can usually be found in the writings of mainstream leftist intellectuals. See: Tony Judt, *Postwar. A History of Europe since 1945* (London: Vintage Books, 2010), 471.

30 *Dabiq*, issue 7, 42.

31 Simon Cottee, "'What ISIS Really Wants' Revisited: Religion Matters in Jihadist Violence, but How?" *Studies in Conflict & Terrorism* 40, no. 6 (June 3, 2017): 448.

32 Nicolas Michael Teausant, from Christian family with military traditions.

33 Jahangir E. Arasli, "Violent Converts to Islam: Growing Cluster and Rising Trend," *GlobalECCO*, accessed June 3, 2018, https://globalecco.org/ctx-v1n1/violent-converts-to-islam.

34 Tony Judt, ibid, 470.

35 Paul Tillich, *Dynamics of Faith* (New York: HarperCollins, 2001).

36 For the elaboration on religious aspects of Europe in crisis see John Gray, "What Scares the New Atheists," *The Guardian*, March 3, 2015, accessed March 28, 2019, www.theguardian.com/world/2015/mar/03/what-scares-the-new-atheists.
37 Personality crisis and frustration are possibly the most frequent global common denominators between ISIS supporters: see examples of Sally-Anne Jones, a 46-year-old former punk musician from Chatham (UK) married a computer hacker and went to Syria with him and second-generation children of immigrants from lower social strata, often involved in petty crime.
38 "Strikingly frequent stories about the corpses of British jihadis bearing tattoos of English football clubs suggest unsuccessful attempts to resolve these [identity] issues." Owen Bennett-Jones, "'We' and 'You'," *London Review of Books*, 27 August 2015, 10.
39 *Foreign Fighters: An Updated Assessment…*, ibid.
40 Ted Robert Gurr, *Why Men Rebel* (Boulder, CO: Paradigm Publishers, 2010).
41 Ibid.
42 "Arab Human Development Report 2009," *UNDP*, 2009, 11, www.undp.org/content/undp/en/home/librarypage/hdr/arab_human_developmentreport2009.html.
43 "Arab Human Development Report 2009," 10.
44 Luis Doncel, "Today's Spanish Youths Are Worse off than a Decade Ago," *ElPais*, November 29, 2018, accessed March 28, 2019, https://elpais.com/elpais/2018/11/28/inenglish/1543402455_039592.html.
45 Colin P. Kelley et al., "Climate Change in the Fertile Crescent and Implications of the Recent Syrian Drought," *Proceedings of the National Academy of Sciences* 112, no. 11 (March 17, 2015): 3241–46.
46 Simon Parkin, "How Isis Hijacked Pop Culture, from Hollywood to Video Games," *The Guardian*, January 29, 2016, accessed March 28, 2019, www.theguardian.com/world/2016/jan/29/how-isis-hijacked-pop-culture-from-hollywood-to-video-games.

Bibliography

06PARIS6995. "Diplomatic Cable." Accessed March 28, 2019. https://wikileaks.org/plusd/cables/06PARIS6995_a.html.

"Abu Bakr al-Baghdadi's call." *Dabiq*, 1. Accessed July 12, 2017. http://media.clarionproject.org/files/09-2014/isis-isil-islamic-state-magazine-Issue-1-the-return-of-khilafah.pdf.

"التصويت : هل تعتبر تقدم تنظيم الدولة الإسلامية في العراق وسوريا لصالح المنطقة؟". [At-Taswit: Hal ta'tabir taqaddum tanzim ad-dawla al-islamiyya fil-Iraq wa-Suriya li-salih al-mintaqa? Survey: Do You Consider the Advancement of the Organization Islamic State to be in the Interest of the Region?] *Al Jazeera*. Accessed 3 June 2018. www.aljazeera.net/votes/pages?voteid=5270.

Al-Wasl, Zaman. "Leaked ISIS Documents Reveal Most Recruits Know Little on Islam." *Haaretz*, August 16, 2016. Accessed March 28, 2019. www.haaretz.com/middle-east-news/leaked-isis-documents-reveal-recruits-ignorant-on-islam-1.5424990.

"Arab Human Development Report 2009." *UNDP*, 2009. Accessed March 28, 2019. www.undp.org/content/undp/en/home/librarypage/hdr/arab_human_developmentreport2009.html.
Arasli, Jahangir E. "Violent Converts to Islam: Growing Cluster and Rising Trend." *GlobalECCO*. Accessed June 3, 2018. https://globalecco.org/ctx-v1n1/violent-converts-to-islam.
Bennett-Jones, Owen. "'We' and 'You'." *London Review of Books*, August 27, 2015.
Brachman, Jarret M. *Global Jihadism: Theory and Practice*. London; New York: Routledge, 2008.
Cottee, Simon. "'What ISIS Really Wants' Revisited: Religion Matters in Jihadist Violence, but How?" *Studies in Conflict & Terrorism* 40, no. 6 (3 June 2017): 439–54.
Doncel, Luis. "Today's Spanish Youths Are Worse Off Than a Decade Ago." *ElPais*, November 29, 2018. Accessed March 28, 2019. https://elpais.com/elpais/2018/11/28/inenglish/1543402455_039592.html.
El Difraoui, Asiem. "Authoritarianism and Radicalization towards Violent Extremism." In *Euromed Survey. Violent Extremism in the Euro-Mediterranean Region*, edited by Josep Ferré, 8th edition, 34–39. Barcelona: IEMed, 2017.
Elgot, Jessica. "Islamic State Actually Has More Support In Britain Than In Arab Nations." *HuffPost UK*, October 16, 2014. Accessed March 28, 2019. www.huffingtonpost.co.uk/2014/10/16/islamic-state-arab-nations-britain-support_n_5995548.html.
"European Union Terrorism Situation and Trend Report 2018 (TESAT 2018)." *Europol*. Accessed March 28, 2019. www.europol.europa.eu/activities-services/main-reports/european-union-terrorism-situation-and-trend-report-2018-tesat-2018.
Fanon, Frantz. *The Wretched of the Earth*. New York: Grove Press, 2007.
Foreign Fighters: An Updated Assessment of the Flow of Foreign Fighters to Syria and Iraq. The Saufan Group, December 2015. Accessed 28 March 2019. http://soufangroup.com/wp-content/uploads/2015/12/TSG_ForeignFightersUpdate_FINAL.pdf.
Gellner, Ernest. *Nations and Nationalisms*. Ithaca, NY: Cornell University Press, 2006.
———. *Postmodernism, Reason and Religion*. London; New York: Routledge, 1992.
Grant, Madeline. "16% of French Citizens Support ISIS, Poll Finds." *Newsweek*, August 26, 2014. Accessed March 28, 2019. www.newsweek.com/16-french-citizens-support-isis-poll-finds-266795.
Gray, John. "What Scares the New Atheists." *The Guardian*, March 3, 2015. Accessed March 28, 2019. www.theguardian.com/world/2015/mar/03/what-scares-the-new-atheists.
Gurr, Ted Robert. *Why Men Rebel*. Boulder, CO: Paradigm Publishers, 2010.
Haupt, Heinz-Gerhard, and Klaus Weinhauer. "Terrorism and the State." In *Political Violence in Twentieth Century Europe*, edited by Donald Bloxham and Robert Gerwarth. Cambridge: Cambridge University Press, 2011.

Hetherington, Marc, and Elizabeth Suhay. "Authoritarianism, Threat, and Americans' Support for the War on Terror." *American Journal of Political Science* 55, no. 3 (July 2011): 546–60.

"Journey of Young Africans into Violent Extremism Marked by Poverty and Deprivation." *UNDP*. Accessed March 28, 2019. www.undp.org/content/undp/en/home/presscenter/pressreleases/2017/09/07/vers-l-extremisme-violent-en-afrique.html.

Judt, Tony. *Postwar. A History of Europe Since 1945*. London: Vintage Books, 2010.

Kelley, Colin P., Shahrzad Mohtadi, Mark A. Cane, Richard Seager, and Yochanan Kushnir. "Climate Change in the Fertile Crescent and Implications of the Recent Syrian Drought." *Proceedings of the National Academy of Sciences* 112, no. 11 (March 17, 2015): 3241–46.

Khosrokhavar, Farhad. *Inside Jihadism: Understanding Jihadi Movements Worldwide*. Abington, Oxon: Routledge, 2015.

Malik, Shiv. "Support for Isis Stronger in Arabic Social Media in Europe than in Syria." *The Guardian*, November 28, 2014, Accessed March 28, 2019. www.theguardian.com/world/2014/nov/28/support-isis-stronger-arabic-social-media-europe-us-than-syria.

Olesen, Thomas, and Farhad Khosrokhavar. *Islamism as Social Movement*. Centre for Studies in Islamism and Radicalisation (CIR), Aarhus University, 2009. Accessed March 28, 2019. www.ps.au.dk/fileadmin/site_files/filer_statskundskab/subsites/cir/pdf-filer/Hæfte2final.pdf.

Ottaway, Marina. *ISIS – Many Faces, Different Battles*. Wilson Center, 2015. Accessed March 28, 2019. www.wilsoncenter.org/sites/default/files/isis_many_faces_different_battles.pdf.

Parkin, Simon. "How Isis Hijacked Pop Culture, from Hollywood to Video Games." *The Guardian*, January 29, 2016. Accessed March 28, 2019. www.theguardian.com/world/2016/jan/29/how-isis-hijacked-pop-culture-from-hollywood-to-video-games.

Pollock, David. "ISIS Has Almost No Popular Support in Egypt, Saudi Arabia, or Lebanon." *WINEP*, October 14, 2014. Accessed March 28, 2019. www.washingtoninstitute.org/policy-analysis/view/isis-has-almost-no-popular-support-in-egypt-saudi-arabia-or-lebanon.

Roy, Olivier. *Globalized Islam. The Search for a New Ummah*. New York: Columbia University Press, 2006.

———. "Le djihadisme est une révolte générationnelle et nihiliste." *Le Monde.fr*, November 24, 2015. Accessed March 28, 2019. www.lemonde.fr/idees/article/2015/11/24/le-djihadisme-une-revolte-generationnelle-et-nihiliste_4815992_3232.html.

———. "Who Are the New Jihadis?" *The Guardian*, April 13, 2017. Accessed March 28, 2019. www.theguardian.com/news/2017/apr/13/who-are-the-new-jihadis.

Roy, Olivier, and Haoues Seniguer, "Comment l'islam est devenu la nouvelle idéologie des damnés de la planète." *Atlantico*, July 4, 2015. Accessed 28 March 2019, www.atlantico.fr/decryptage/comment-islam-est-devenu-nouvelle-ideologie-damnes-planete-olivier-roy-haoues-seniguer-2221200.html.

"San Joaquin County Man Pleads Guilty to Attempting to Provide Material Support to ISIL." *Press Release*. Federal Bureau of Investigation, 1 December 2015. Accessed 28 March 2019. www.fbi.gov/contact-us/field-offices/sacramento/news/press-releases/san-joaquin-county-man-pleads-guilty-to-attempting-to-provide-material-support-to-isil.
Schuster, Courtney, David Sterman, and Peter Bergen. "ISIS in the West. The New Faces of Extremism." *New America Foundation*, November 15, 2015. Accessed March 28, 2019. www.newamerica.org/international-security/future-war/policy-papers/isis-in-the-west/.
Temple-Raston, Dina. "New Terrorism Adviser Takes A 'Broad Tent' Approach." *NPR.org*, January 24, 2011. Accessed March 28, 2019. www.npr.org/2011/01/24/133125267/new-terrorism-adviser-takes-a-broad-tent-approach.
Tillich, Paul. *Dynamics of Faith*. New York: HarperCollins, 2001.
Veldhuis, Tinka, and Jørgen Staun. *Islamist Radicalisation: A Root Cause Model*. The Hague: Netherlands Institute of International Relations Clingendael, 2009.
Venhaus, John M. "Why Youth Join Al-Qaeda." United States Institute of Peace, May 2010. Accessed March 28, 2019. www.usip.org/publications/2010/05/why-youth-join-al-qaeda.
Wintrobe, Ronald. "Extremism, Suicide Terror, and Authoritarianism." *Public Choice* 128, no. 1/2 (2006): 169–95.

Conclusion

As this book goes to print three events coincide. In March 2019 ISIS was declared defeated by the Kurdish forces in Syria. The victory was only partial – it put an end to the territorial entity called the caliphate but thousands of ISIS supporters, including the command, fled Syria and Iraq with hundreds of millions of dollars when their defeat was looming. With or without the so-called caliphate jihadism as an ideology remains vibrant – it may be a matter of time only that the world will be introduced to its next emanation.[1] The likelihood of such a scenario is corroborated by yet another wave of state repression and violence that is sweeping the Middle East.[2] Simultaneously in Algeria, Abdelaziz Bouteflika – the fifth president after independence in 1962, in office for exactly 30 years now – is on the verge of being ousted from it. The transition may be controlled without depriving the authoritarian state of any power but the ouster of Bouteflika was caused by mass student protests, demanding "state governed by justice and the rule of law."[3] In a region where counterrevolutionary forces have had the upper hand after the 2011 revolts Algeria again gives the south hope. The defeat of ISIS and Bouteflika's ouster coincide with the tragedy in Christchurch in New Zealand, where a right-wing extremist killed 50 Muslims in prayer in March 2019. The terrorist funded European far-right racist organisations in Austria and travelled around Europe before committing the crime. Christchurch might as well have happened on the European soil.

All three events show how perpetually actual is the issue of political violence and the north/south divide. This study has revisited these themes with the aim of adopting a global perspective or at least a less Eurocentric one. It has combined a seminal voice on political violence in the south (Fanon) with that of the north (Arendt) to find that their tones were not dissimilar. In the course of analysis, it became evident that both philosophers identified similar political problems in the south and north. These problems have not lessened since the 1960s and political violence (here analysed in the form of jihadism) is a response to them. The overall conclusions are threefold: about (1) commonalities between Fanon and Arendt, (2) the interpretation of jihadi violence today, (3) the political sphere in a globalised world.

1.

Fanon's and Arendt's work and subsequent debate about political violence prove impervious to time, not least because "the structures and issues involved remain constant,"[4] which they indeed do. As was shown in Chapter 2, their philosophies of violence require a finer contextual reading, which ultimately reveals that neither did Fanon advocate unconditional violence, nor was Arendt so unconditionally against all forms of violence. It may be that Arendt's underdeveloped stance on racial and north-south issues, or her blindness to "structural and epistemic violence of colonialism" – as Hanssen has it[5] – disproportionately widens the disagreement with Fanon on violence, power and agency where there would be little disagreement otherwise. In one of early reviews of this book it was suggested that in order to bring Fanon and Arendt together two methods could be applied: either deradicalising Fanon or radicalising Arendt. In fact, the way Fanon's and Arendt's thoughts on violence are read here shows that both methods can be applied simultaneously (and with lesser injustice to the existing literature than if only one was at work). When it comes to "radicalisation", Arendt and Fanon meet half way.

Both, Fanon and Arendt, share the method: they both take a psychological and socio-political position – both are outsiders hailing from the oppressor and the oppressed at the same time. Their philosophies are a product of the "European evil" (the Second World War and colonialism), and both are philosophers of beginnings prescribing a grandiose humanist project. As demonstrated in Chapter 3, from the European fall sprouts Arendtian and Fanonian search for the new humanism. Repulsed by the enormity of violence they both dream of a new political man. Arendt finds him in the north, in the public space that the Greeks infused into European politics. Fanon is unable to find him in the overwhelmingly racist north so he puts his hope in the south. Both are philosophers of responsibility and commitment and both see human act (Fanon) or action (Arendt) as a necessary vehicle of a man creating himself. Both see a nation as formed by bonds of joint action rather than built on xenophobic community of blood. Their non-xenophobic nationalism is anti-racist and anti-tribal. They call for the same true representation of reality in words and new relations in deeds, although both add an emotional, religious or mystical element to politics.

Both in fact can be interpreted as philosophers of north-south unity in their own right. It may seem an overstatement with regards to Arendt in particular as her thought would sometimes be seen as racist or ignorant of the south but unwittingly she is the one that preaches a revival or renewal of a political creature that the southerner can be. If she praised the Czechoslovaks in 1968 as the most political, and, hence, powerful creatures she would praise the Arab awakening of 2011 when peacefully millions of Egyptians, Tunisians and Syrians demonstrated their ability for collective political action. Fanon is radically empathic towards the dispossessed, but Arendt, too, is compassionate and shows commitment to being the other, albeit not explicitly in OV.

In BPF Arendt asserts, openly and in complete agreement with Fanon, that public space and freedom are the fruit of the struggle against occupation, yet she refuses to assert the same opinion in relation to anti-colonial struggle. Elsewhere, Arendt herself is of a view that only true liberation of the people (of the south) will make them a polity. Both Arendt and Fanon share criticism of Western culture, which is dangerously walking towards the edge. The more of the (technically) impossible we are cable of achieving the less of the possible we can.

Fanon, with his claim of being both the slave of his appearance and the slave of the coloniser stretches Arendt's concept of politics twofold. Firstly, by showing that there is a mask that never comes off: black skin. And secondly that nonviolence or value-based revolutionary student movements are impossible in the Third World because of the non-existence of public space.

2.

Fanon did not and most likely would not espouse jihadist ideology. Religious tension can be a form of racism, he asserted, especially that Africa is racially divided into the white north and black Sub-Saharan south. What he called "mythical reality" – the product of religion – was a primary obstacle to freedom and liberation. Nationalism, too, easily turns to tribalism. Yet he saw immediate benefits if the struggle for national liberation was linked to a "cultural phenomenon," such as the awakening of Islam. Jihadist violence stands against most of what Fanon advocated: humanism, universality, commitment and responsibility.

Table C.1 Conditions for violence in the global south and north, identified by Arendt and Fanon

South	*North*
Lack of values and cultural void	Culture of selfishness
Disunity, lack of solidarity, part of the colonised turned colonists	
Dissolution of responsibility – hypocrisy	
Weak and inactive elite	Tired and inactive elite
State violence: military coups, corruption, exploitation, terrorism	Rule by nobody, bureaucratisation
None or little developed public space, little possibility of joint action	Decline of party systems
Strong mythical reality (religion, customs) – religious xenophobia	False promise of a positivist progress – resentment of bigness (globalisation)
Tribalism	
Class societies: inequalities and disconnected social strata	
Race relations: biologisms	
Weak education	

The analysis of the reasons for jihadist appeal (Chapter 4) reveals that the faults of the socio-political systems in the south and north, as identified by Fanon and Arendt, have not lessened. Fanon and Arendt identify the conditions that push people to irrational, disorganised, destructive violence (Table C.1). Jihadist ideology, in light of the persistence of these problems, becomes a favourable and vindicating programme, mostly to southerners but increasingly in the north as well.

3.

The decolonisation, the struggle for freedom was to be an abrupt, violent, man-forming period on the ashes of which the phoenix of the new man and the new humanity would rise. Little did Fanon know about the long decolonisation that ensued. He sensed it in the assimilation to the metropolis of the Third World elite and the undecidedness of the newly independent peoples. He sensed it in the poverty that was not going away – no reparations were ever paid. The more he wrote WotE the little change, of the kind he had hoped for, transpired. Decolonisation was not over. Neither is it over more than half a century later. A social feeling that persists today in the Middle East and the Arab world is that of indignity and of being robbed.[6] Still, as Fanon observed "enormous wealth rubs shoulders with abject poverty" (WoTE, 116) in the south – in Zamalek, the richer and most liberal part of

Cairo sips drinks in the Marriott that overlooks a shanty town across the river. Tired, swollen, old hands of a 10-year-old in an Egyptian village called Izbet Bahariya contrasts with pristine cleanness of the American University in Cairo campus. Continuously, rage has had ripe circumstances to grow in the south.

Likewise in the north, even if less ostentatiously. As Arendt warned the power to act would shrink still. As political beings we are in the need of a new example if we are not to become chicken or rats ruled by an elite who thinks that "people in think tanks are thinkers and computers can think" (OV, 83). People are naturally hungry for action – if they cannot act politically, they are kindled by violence. It directly explains why terrorism and violence grows in the Middle East where the public sphere is limited and in the north, where social atomisation limits political action. Yet, for some, the transposition of Arendt's socio-political critique to contemporary relations is not straightforward. Just like Fanon could have been seen as a preacher of jihadist movements, she may as well be seen as a preacher of alt-right or other populist movements that nominally champion "direct" democracy, immediately giving people the voice and possibility for action. They oppose established democratic institutions, want to dismantle them on the premises of doing away with the elite, while these institutions are the ultimate guarantors of consensual coexistence and democratic legal orders against the return of non-democratic tyranny of majority. Truly people are hungry for action but democratic institutions seem to capitulate in front of the people's urge to act; submitting to dictates of majority rather than instilling participatory democracies. Politics, a sphere of immortality and miraculous concerted action, has become a regular job, void of vision. Fear is emphasised and promoted at the expense of hope, and ultimately the allure of violence grows because, among other innuendos, it gives a semblance of collective action.

On top of these problems comes a binary and unyielding division between the figure of a European and the figure of an Arab or Muslim. That division is instilled by all the worst and most petrifying trends in foreign and domestic policies of European and Arab countries. Some want to see this division as formative. Leszek Kołakowski went as far as to attribute the birth of European identity to the historical conflict with the Muslim world.[7] Racial relations (European-Arab are seen as such too) are particularly prone to biological metaphors (the colour of the skin is a natural fact) and, hence, to violence. Racism is an ideology based on pseudo-scientific theories, but race is a fact (OV, 75). Arendt would warn that black riots and potential white backlash are "not yet manifestations of racist ideologies" but, if they escalate, they could trigger racist ideologies to justify the growing violence. Mere prejudices turn into ideologies. The same argumentation can be built with regards to violence attributed to Muslims: the more jihadist violence, the greater the backlash and the easier these relations will turn into racist ideologies. Politics of fear, rather than hope build on these ideologies, augmenting them, in turn. The message we receive is that differences between Europeans and Muslims/Arabs are simply insurmountable. When this conflict becomes a conflict of ideologies then "law and order would indeed become a façade" (OV, 76). The majority would be willing to live in a police state. Arendt warns of a scenario that is possible today – Arabs and Muslims, physically different (veil, brown skin, clothing) and seen as responsible for the cultural background to the most potent contemporary terrorist movements – invite racism and violence, which, for now, is mostly dormant. Christchurch attacks forewarn of what there may be to come. In *A History of Bombing* – published at the height of the post-cold war optimism – Sven Lindquist ominously reminds:

> Out of this violence both that which has already been committed and that which is still dormant, the century's dreams of genocide emerge. The injustice we defend forces us to hold on to genocidal weapons with which our fantasies can be realized whenever we like. Global violence is the hard core of our existence.[8]

Arendt's polemics with Fanon about violence and power, their diagnosis of socio-political relations in the north and south teach that the political sphere is shrinking globally, despite greater technical connectivity between people. "Every decrease in power is an open invitation to violence" (OV, 86). Global inequalities were not a topic of this book, but it does not take an unusual power of imagination to see how they can add to the feeling of injustice or to the fear of losing privileges. Governments and elected politicians fall hostage to their own narratives and prove unable (and/or unwilling) to lead with hope and a message of reconciliation between the south and the north. The great power of politics, recognised by Fanon and Arendt, is that it is possibly the most potent ideological trend-setter. Yet it is again setting the trend against a gigantic ocean of more than 1.5 billion people who have become a substitute culprit for northern misfortunes. Unless this is reversible the hope of humanity lies indeed in the southerner, who – despite far more adverse conditions to his agency – will make the insane effort of rebelling against the north-south dichotomy. The first editor of WotE, perhaps prophetically, picked the following quotation from the last page of Fanon's text for the cover and squeezed it between hopeful green lines: "Il s'agit pour le Tiers-Monde de recommencer une histoire de l'homme [It is for the Third World to restart the history of man]."[9]

Notes

1 Fawaz Gerges, "The Islamic State Has Not Been Defeated," *The New York Times*, March 24, 2019, accessed March 28, 2019, www.nytimes.com/2019/03/23/opinion/isis-defeated.html.
2 In March 2019 Egypt executed 15 men convicted in unfair trials. Ruth Michaelson, "Nine Executed in Egypt over Hisham Barakat Assassination," *The Guardian*, February 20, 2019, accessed March 28, 2019, www.theguardian.com/world/2019/feb/20/egypt-executes-nine-men-convicted-hisham-barakat-assassination.
3 Anthony Dworkin, "Algeria's Protests: A View from the Ground," *ECFR*, March 25, 2019, accessed March 28, 2019, www.ecfr.eu/article/commentary_algerias_protests_a_view_from_the_ground.
4 David Macey, *Frantz Fanon: A Biography* (London; New York: Verso, 2012), 496.
5 Jens Hanssen, "Translating Revolution: Hannah Arendt in Arab Political Culture," *Journal for Political Thinking HannaArendt.net* 7, no. 1 (2013), accessed March 28, 2019. www.hannaharendt.net/index.php/han/article/view/301.
6 Hazem Fahmy, "An Initial Perspective on 'The Winter of Discontent': The Root Causes of the Egyptian Revolution," *Social Research* 79, no. 2 (Summer 2012): 349–76.
7 Leszek Kołakowski, "Szukanie barbarzyńcy. Złudzenia uniwersalizmu kulturowego," in *Czy diabeł może być zbawiony i 27 innych kazań* (Kraków: Znak, 2012), accessed March 28, 2019, www.znak.com.pl/kartoteka,ksiazka,3437,Czy-diabel-moze-byc-zbawiony.
8 Sven Lindqvist, *A History of Bombing*, trans. Linda Haverty Rugg (New York: The New Press, 2001), 186.
9 Frantz Fanon, *Les Damnés de La Terre*, Cahiers Libres, Nos 27–28 (Paris: Éditions Maspero, 1962).

Bibliography

Arendt, Hannah. *On Violence*. Orlando; Austin; New York; San Diego; London: HMH, 1970.

Dworkin, Anthony. "Algeria's Protests: A View from the Ground." *ECFR*, March 25, 2019. Accessed March 28, 2019. www.ecfr.eu/article/commentary_algerias_protests_a_view_from_the_ground.

Fahmy, Hazem. "An Initial Perspective on 'The Winter of Discontent': The Root Causes of the Egyptian Revolution." *Social Research* 79, no. 2 (Summer 2012): 349–76.

Fanon, Frantz. *Les Damnés de La Terre*. Cahiers Libres, Nos 27–28. Paris: Éditions Maspero, 1962.

———. *The Wretched of the Earth*. New York: Grove Press, 2007.

Gerges, Fawaz. "The Islamic State Has Not Been Defeated." *The New York Times*, March 24, 2019. Accessed March 28, 2019. www.nytimes.com/2019/03/23/opinion/isis-defeated.html.

Hanssen, Jens. "Translating Revolution: Hannah Arendt in Arab Political Culture." *Journal for Political Thinking HannaArendt.net* 7, no. 1 (2013). Accessed March 28, 2019. www.hannaharendt.net/index.php/han/article/view/301.

Kołakowski, Leszek. "Szukanie barbarzyńcy. Złudzenia uniwersalizmu kulturowego." In *Czy diabeł może być zbawiony i 27 innych kazań*. Kraków: Znak, 2012. Accessed March 28, 2019. www.znak.com.pl/kartoteka,ksiazka,3437,Czy-diabel-moze-byc-zbawiony.

Lindqvist, Sven. *A History of Bombing*. Translated by Linda Haverty Rugg. New York: The New Press, 2001.

Macey, David. *Frantz Fanon: A Biography*. London; New York: Verso, 2012.

Michaelson, Ruth. "Nine Executed in Egypt over Hisham Barakat Assassination." *The Guardian*, February 20, 2019. Accessed March 28, 2019. www.theguardian.com/world/2019/feb/20/egypt-executes-nine-men-convicted-hisham-barakat-assassination.

Index

Note: **Bold** page numbers refer to tables.

Milton Keynes UK
Ingram Content Group UK Ltd.
UKHW022358061024
449327UK00031B/2563

9 780367 787905